Johannes Zachhuber
Time and History in Denis Pétau

CHRONOI
Zeit, Zeitempfinden, Zeitordnungen
Time, Time Awareness, Time Management

―――

Edited by
Eva Cancik-Kirschbaum, Christoph Markschies and Hermann Parzinger

on behalf of the Einstein Center Chronoi

Volume 17

Johannes Zachhuber

Time and History in Denis Pétau

Philosophy, Science, and Religion in Early Modern France

DE GRUYTER

ISBN 978-3-11-914203-8
e-ISBN (PDF) 978-3-11-222334-5
e-ISBN (EPUB) 978-3-11-222373-4
ISSN 2701-1453
DOI https://doi.org/10.1515/9783112223345

This work is licensed under the Creative Commons Attribution-NonCommercial-NoDerivatives 4.0 International License. For details go to https://creativecommons.org/licenses/by-nc-nd/4.0.

Library of Congress Control Number: 2025945497

Bibliographic information published by the Deutsche Nationalbibliothek
The Deutsche Nationalbibliothek lists this publication in the Deutsche Nationalbibliografie; detailed bibliographic data are available on the internet at http://dnb.dnb.de.

© 2026 the author(s), published by Walter de Gruyter GmbH, Berlin/Boston, Genthiner Straße 13, 10785 Berlin. This book is published with open access at www.degruyterbrill.com.

www.degruyterbrill.com
Questions about General Product Safety Regulation:
productsafety@degruyterbrill.com

For Christian Richard Gerhard Zachhuber
* 18 September 2023

So bleibe denn die Sonne mir im Rücken!
Der Wassersturz, das Felsenriff durchbrausend,
Ihn schau' ich an mit wachsendem Entzücken.
Von Sturz zu Sturzen wälzt er jetzt in tausend
Dann aber tausend Strömen sich ergießend,
Hoch in die Lüfte Schaum an Schäume sausend.
Allein wie herrlich diesem Sturm ersprießend,
Wölbt sich des bunten Bogens Wechsel-Dauer,
Bald rein gezeichnet, bald in Luft zerfließend,
Umher verbreitend duftig kühle Schauer.
Der spiegelt ab das menschliche Bestreben.
Ihm sinne nach und du begreifst genauer:
Am farbigen Abglanz haben wir das Leben.
Goethe, *Faust. Der Tragödie zweiter Teil.*

Let the sun shine on, behind me, then!
The waterfall that splits the cliffs' broad edge,
I gaze at with a growing pleasure, when
A thousand torrents plunge from ledge to ledge,
And still a thousand more pour down that stair,
Spraying the bright foam skywards from their beds.
And in lone splendour, through the tumult there,
The rainbow's arch of colour, bending brightly,
Is clearly marked, and then dissolved in air,
Around it the cool showers, falling lightly.
There the efforts of mankind they mirror.
Reflect on it, you'll understand precisely:
We live our life amongst refracted colour.
Goethe, *Faust. Der Tragödie zweiter Teil.*
Translation: A. S. Kline

Acknowledgments

This book is the result of an eight-month fellowship at the Einstein Center Chronoi in Berlin, which I held in the academic year 2023/24. My interest in Denis Pétau had previously been piqued as part of my research into the reception of early Christian philosophy and the historicization of Christian thought. In this connection, I came to know Pétau as a key player in the emergence of modern historical theology. But while I was aware of his role in debates about early modern scientific chronology, it is doubtful that without the opportunity offered by the Chronoi Center I would have discovered his *De doctrina temporum* as the resource it is for ideas about time and history – ideas which ultimately underlie his approach to theology as well. I am therefore truly grateful to the Center, its directors and staff, for providing me with the research time I needed for the exploration of this fascinating topic.

The Chronoi is an amazing institution. Its thematic focus on the study of time provides the fellowship with a common intellectual concern, but its broad, interdisciplinary approach to the topic also makes it intellectually diverse. The relatively small size means that fellows feel like being part of a family. The personable and welcoming atmosphere is largely due to the wonderful admin team led by Stefanie Rabe on whose support in a variety of matters I could always rely.

Early results of my research were presented at the Chronoi's own research seminar where the fellows present offered helpful comments. Later, I spoke about my findings at the biennial Oxford-Bonn seminar. Eva Cancik-Kirschbaum read the whole manuscript and offered valuable feedback. Ivan Parga Ornelas cast the expert eye of the Neolatinist on the entire text; I am enormously grateful for his careful scrutiny and correction of the Latin citations and my translations. Remaining mistakes are, of course, entirely my own responsibility.

A special highlight of my year in Berlin were the regular conversations I enjoyed with Christoph Markschies who made time for me despite his incredibly busy schedule.

As ever, I could not have done my work without my wife, Lydia Schumacher. Our daily companionship sustains me in more ways than I can express. She held a fellowship at the Chronoi almost coterminous with mine, so we spent a year together in Berlin. As it happened, this year became unique for us for reasons entirely unrelated to our research. During our stay, on 18 September 2023, our son, Christian Richard Gerhard Zachhuber, was born. Since then, he has brought us immense joy and continues to do so on every single day. To him this book is dedicated.

Oxford, July 2025
Johannes Zachhuber

Contents

Acknowledgments —— IX

1 Introduction —— 1
1.1 Subjective time as social reality —— 1
1.2 Denis Pétau and his theory of social time —— 3
1.3 Pétau and early modern historiography —— 9
1.4 The scope of this study —— 12

2 Denis Pétau: A scholar between tradition and modernity —— 14
2.1 Life —— 14
2.2 Works —— 18

3 *De doctrina temporum* in the context of early modern chronology —— 24
3.1 Joseph Scaliger and scientific chronology —— 24
3.2 Pétau's *De doctrina temporum* —— 32

4 Pétau's theory of time —— 38
4.1 The definition of time —— 40
4.2 The Aristotelian background —— 42
4.2.1 One time or many? —— 42
4.2.2 Material and formal time —— 46
4.3 Subjective time as social time —— 49
4.4 A new science —— 50
4.5 Conclusion —— 54

5 Pétau on history —— 56
5.1 History as distinct from chronology —— 56
5.2 History as the realm of factuality —— 61
5.3 Three methodological principles —— 67
5.4 Chronology, time, and the B.C./A.D. system —— 78
5.5 Conclusion —— 84

6 Social time as religious time —— 86

7 Time and God —— 93

8 Conclusion —— 102

Bibliography —— 107

General Index —— 117

Index of Passages —— 120

1 Introduction

1.1 Subjective time as social reality

Questions about time can emerge in different contexts. Throughout Western intellectual history, two such contexts have been extensively studied. On the one hand, there is cosmic time, a seemingly objective, irreducible element of the universe in its totality as well as its parts. It is therefore studied by the science of nature which Aristotle called physics. On the other hand, there is time as subjectively experienced by human beings. As it arises within human consciousness, it must be investigated by psychology or the philosophy of mind. Quite how these two aspects of time are related has been controversial since antiquity and, in a sense, remains debated into our own time.[1] It would perhaps seem obvious that any subjective perception of time must presuppose the existence of objective time, unless time is deemed an illusion. But unlike other 'objects' in the natural world, time is not an observable entity. All things exist in time, but it apparently needs a timekeeper to establish this fact about the world. No time, then, without someone's awareness of time.

Attempts have therefore not been lacking to prioritise the subjective dimension of time. This may already have been the position of St Augustine when he defined time as the 'distension of the soul' meaning that time is prefigured in the structure of human consciousness.[2] The church father also believed that God made time as part of his creation of the cosmos; he thus probably did not think time was primarily a matter of human awareness of time.[3] Later thinkers, however, pursued this approach, and in the twentieth century, it was emphatically advanced by Martin Heidegger for whom time was primarily a human 'existential'.[4]

For Heidegger and those influenced by him, affirming the primacy of subjective time meant inscribing time into human existence. Being who we are, we cannot but think of ourselves as stretched from a past through the present and into

---

[1] On the ancient debate see Zachhuber, *Time and Soul*. For its relation to contemporary issues see Detel, *Subjektive und objektive Zeit*. Perhaps the most important recent contribution is Ricoeur, *Temps et récit*.
[2] Augustine, *Confessiones* XI 26, 33: *inde mihi visum est nihil esse aliud tempus quam distentionem; sed cuius rei, nescio, et mirum si non ipsius animi*. On this theory see: Corti, *Zeitproblematik*.
[3] Augustine, *Confessions* XI 14, 17.
[4] Heidegger, *Sein und Zeit*.

the future. It is this existential structure, Heidegger argues, which only explains our concern for time in the world around us including at the cosmic level.

One can accept Heidegger's claim about the existential temporality of human beings without swallowing whole his far-reaching ontological assumptions. Existing consciously in time would then appear as a foundational determination of humanity. As human beings, we inevitably have a concern for time. We are aware that we have a past and look forward to a future, and this awareness defines us as who we are. We do not simply live in the present moment but know that we are the products of our history and that our current actions will have consequences for the time to come.

On a more practical level, this means that human beings are by nature time reckoning. It is arguable that we cannot conceptualise our temporality without applying units of time measurement to it. Expressed differently, we necessarily express our temporality by imposing a temporal structure on the world of our experience. We distinguish days and nights, hours within the day. We count time in weeks, months, and years. In this way, we structure our own lives and relate to past and future.

Importantly, however, we do not do so on our own. Timekeeping is by necessity a communal undertaking. This not only in the sense that we need help in orienting ourselves in time. More fundamentally, subjective time is a shared language. It is only helpful to us to the extent that we can communicate about it with others; and to that end, it needs to be intelligible to those with whom we communicate. If a friend asks me where I was two weeks ago, it is essential that we both share a common understanding of a 'week'.

Timekeeping is thus based on a social consensus concerning units of time as well as their modes of measurement. This social dimension comes to the fore even more strongly once we move from the kind of practical issue considered above to ways in which the synchronicity of time-keeping matters for societies as a whole. If time is a dimension of individual human existence, as Heidegger maintained, the same is emphatically the case for human communities and societies. Across cultures and throughout history, human societies have defined themselves in temporal terms.[5]

One notable aspect of this reality are regular holidays and festivals which are celebrated jointly by members of the community on a certain day or on certain days. These days have to be determined in such a way that everybody observes them at the same time. Yet once again, this is more than a pragmatic need for a social consensus. Rather, these days are usually related to astronomical phenom-

---

5 Assmann, 'Kulturelle Zeitgestalten'.

ena, such as the new moon or the summer solstice. 'Subjective' time at the societal level is thus directly related to, and in a sense based on, the 'objective' time of the cosmic bodies and their regular movements.

I started from the observation that the problem of time emerges in different contexts. We can now discern at least one additional such context, namely, the study of calendars and their social functions. This study broadly has two purposes: to improve existing calendars and to understand the calendric systems used by societies in the past. It therefore involves a mix of disciplines. It cannot dispense of mathematics and astronomy, but it also depends on the careful scrutiny of historical sources that contain information about the principles on which past societies built their calendric systems. Objective and subjective time thus come together in this approach which understands time as an intersubjective or social phenomenon.

It is a key thesis of the present study that such a theory of social time was proposed as part of the debates about chronology and the study of calendars in early seventeenth century France. Its originator was Denis Pétau, an erudite Jesuit and a keen participant in many of the most burning controversies of the time. While his contribution to the contemporaneous quest for scientific chronology has not been overlooked in previous scholarship, it will be shown in what follows that Pétau remains in many ways an untapped resource for creative ideas about time and history.

1.2 Denis Pétau and his theory of social time

The history of the study of time in the context of calendric time measurement reaches back to the earliest civilisations, but it took on particular urgency in early modernity.[6] At this point, the West had used the so-called Julian calendar for over one and a half millennium. Introduced by Julius Caesar in 45 B.C., this calendar was based on an assumed solar year of 365.25 days. It therefore stipulated normal years of 365 days as well as quadrennial leap years. Yet the solar year is, in fact, somewhat shorter at c. 365.2422 days, and while the difference is initially small, it makes the Julian calendar gain a full day every 129 years. By the sixteenth century, therefore, the date according to the Julian calendar was ten days behind the actual solar cycle. Thus, the spring equinox, would not occur on 21 March but already on the 11th. This mattered not least because the date of

---

[6] For an overview see Richards, *Mapping Time*. For the medieval prehistory see Nothaft, *Scandalous Error*. For the background see also Weichenhan, *"Ergo perit coelum ..."*, 85–102.

Easter was fixed in the Ecclesiastical calendar as occurring on the Sunday following the first full moon after the March equinox. A deviation of the civil calendar from its astronomical basis therefore distorted the dating of this Christian festival.

Calls for calendar improvement thus became louder during the sixteenth century. They were ultimately answered in the Gregorian calendar reform of 1582, named after Pope Gregory XIII (1502–1585).[7] Based on the best available astronomical data, the scholars advising the pope calculated the solar year at 365.2425 days. They therefore proposed to skip leap years where the year was divisible by 100, except where it was also divisible by 400. Thus, three out of four centurial years, which would have been leap years according to the Julian calendar, became normal years of 365 days in the Gregorian calendar. The Gregorian reform also moved the calendar forward by ten days to realign it with the dates underlying the Easter date in the Early Church.

Being spearheaded by the Roman pontiff, the new calendar could not but be controversial in the newly divided world of Western Christendom. Catholic countries adopted the new calendar as soon as it was published, but neither Protestant nor Eastern Orthodox countries initially followed their example. Rather, debates continued for more than a century, and it was not until the mid-eighteenth century that the Gregorian calendar became generally accepted across Western Europe.[8]

At the same time, the scholarly investigation of a wide range of historical sources made urgent the synchronisation of the many different calendric systems in use by their authors. The study of calendars was thus closely connected with a strong contemporaneous interest in history fuelled by the Humanist imperative of a return *ad fontes*. This led to a growing interest in chronology in order to establish correct dates for historical events and express them in ways intelligible to the readership of the time.[9]

A key participant in these various debates was Denis Pétau, a French Jesuit, often also referred to as Dionysius Petavius according to the custom among the literati of the time.[10]

---

7 On the Gregorian reform: Coyne, Hoskin, and Pedersen (eds.), *Gregorian Reform of the Calendar.*
8 For the development in the Holy Roman Empire see Koller, *Strittige Zeiten.* For England cf. Poole, *Time's Alteration.*
9 This topic has found considerable attention in recent decades stimulated above else by Anthony Grafton's work. See e.g. his 'Joseph Scaliger and Historical Chronology' as well as the monumental, two-volume *Joseph Scaliger,* esp. vol. 2: *Historical Chronology.*
10 Pétau has not been served well by recent scholarship. By far the most comprehensive scholarly treatment is Hofmann, *Theologie.* For a review of earlier literature, see Chapter 2 below.

Proudly Catholic, Pétau celebrated the success of papal calendar reform as proof of the superiority of Roman Catholicism.[11] Similarly, his own involvement in the scholarly treatment of chronology was motivated by the need to produce a Catholic counterweight to recent advances in this discipline among Protestants, notably the great Joseph Scaliger.[12] Yet if Pétau embraced the spirit of confessional polemic so characteristic of his age, he was nevertheless also an extremely scrupulous and conscientious scholar of stupendous learning. His contributions to scientific chronology, published initially in his two-volume *De doctrina temporum*, have generally been accepted as definitive and a considerable improvement over the state the discipline had previously attained.[13]

Pétau's ambition, nevertheless, went beyond the detailed comparison of historic calendars and the establishment of an exact chronology. In keeping with the title of his great work, he aimed at nothing less than a 'doctrine of time', a novel and truly scientific understanding of time itself.[14] This ambition has not gone unnoticed in previous literature. Pétau has often been credited with a key contribution to the transition from a 'relative' to an 'absolute' conception of time in early modernity. According to this assessment, Pétau's attempt to synchronise the different calendars he found used in the sources he examined led him to the stipulation that *only one* calendar could in principle be the correct one. Rather than simply comparing the various calendric schemes used across cultures in order to make them mutually intelligible (as shared languages, to use the expression introduced above), Pétau would have advocated for the *replacement* of diversity with a single, unified calendric and chronological system.

The charge was led by Donald Wilcox in his 1987 book, *The Measure of Times Past*. The work argued against the backdrop of the postmodern critique of the tyranny of modernity that a concept of absolute time existed only during the three centuries between the early 1600s and the middle of the twentieth century. Absolute time, of course, is the established name for Newton's specific theory according to which time exists independently of events or bodies; 'of itself, and from

---

11 Petavius, *De doctrina*, vol. 1, *Prolegomena* 2, sig. e5v.
12 According to his biographer, François Oudin, Pétau's father, Jerome, 'lui disoit souvent, qu'il devoit se mettre en état de combattre & de terrasser le *Geant des Allophyles*: C'est ainsi qu'il nommoit *Joseph Scaliger*, don't l'érudition & les Ouvrages donnoient un grand relief au parti Protestant': 'Denis Petau', 83.
13 The fullest treatment of Pétau's chronology is Di Rosa, 'Denis Pétau e la cronologia'.
14 Petavius, *De doctrina*, vol. 1, *Epistola*, sig. a6r: *Vel reipsa vel, ut dicam modestius, opinione nostra, nunc primum in absolutae scientiae redacta formam, quasi nova quaedam disciplina prodeat.*

its own nature, flows equably without relation to anything external'.[15] But Wilcox maintained that an analogous notion of time took hold in historical consciousness at the same point in history. This 'absolute' time of the chronographers was the B.C./A.D. dating system which ...

> [...] displays all of the features of Newtonian time. Indefinitely extended forward and backwards from an arbitrary point, it is truly universal in application and seems to carry with it no thematic or interpretive weight. It can date events which have no intrinsic relationship to one another and quantify their temporal distance precisely and unambiguously. It is a continuous, universal, and endless time line with no external reference, a chronology perfectly adapted to the needs of Newtonian time.[16]

This system, according to Wilcox, was introduced by Pétau, whom he inexplicably calls 'Domenicus' Petavius, 'in 1627, ten years before Descartes proposed in his *Discourse on Method* some of the essential notions behind the new scientific methodology'.[17]

Pétau, it has to be conceded right away, did not in fact 'invent' the B.C./A.D. system although it is arguable that his work was instrumental in popularising it.[18] Wilcox is, however, right in pointing out that Pétau considered the birth of Christ as the key reference point in that system a mere convention all the more since he was aware 'of the discrepancy between the actual birth date of Christ' and the year on which the dating system counted the years.[19]

How did this new regimen of time differ from what was current before? According to Wilcox, 'dating systems in use before Newton were not absolute and did not contain the implications about absolute time that characterise the B.C./A.D. system.'[20] In this sense, time was 'relative', not absolute as dates were 'tied to specific themes, events and moral lessons'.[21] Whereas post Petavius time measurement became always prior to any interpretation of what happened in any given history, the reverse was true before his revolution. Historians *first* assessed the significance of given events or developments, and only *then* did they integrate them into a time frame.[22]

---

15 Newton, *Scholium*, I.
16 Wilcox, *Measure of Times Past*, 7–8.
17 Wilcox, *Measure of Times Past*, 8.
18 Cf. Klempt, *Säkularisierung*, 86–89.
19 Wilcox, *Measure of Times Past*, 8.
20 Wilcox, *Measure of Times Past*, 9.
21 Wilcox, *Measure of Times Past*, 9.
22 Wilcox, *Measure of Times Past*, 9.

Wilcox is not blind to the enormous practical advantages 'absolute' time provides. He acknowledges that the modern dating system is

> ... a truly remarkable instrument and worthy to be the symbol of objectivity and certainty for historical facts. It can perform the most minute and immediate quotidian tasks – organize our checkbooks, fix our entry into school or our retirement from the work force, and calculate fiscal and legal responsibilities. The same dating system can locate such distant events as the battle of Marathon, the period when people first engaged in agriculture, and even the time when life on earth began.[23]

Yet it has its drawbacks, and the condescension with which modern historians look back to their forebears bereft of the blessings of our own chronological tools may be misplaced as to an extent a plurality of timeframes corresponds better to the variegated ways in which human beings and their communities experience and relate to time.

Wilcox' suggestion has been eagerly received by anthropologists and classical historians. In his 2007 Sather lectures, Denis Feeney relies on Wilcox for the claim that 'Domenicus Petavius' was the first to 'expound the B.C./A.D. system as a basis for a universal time line for scholars and historians', to move on to a systematic exploration of the plurality of ancient timeframes and the efforts of synchronisation that this state of affairs necessitated.[24] Feeney's interest overall is in historical understanding. Despite the 'incalculable benefits in terms of convenience and transferability' which the modern dating system has brought, he emphasises, it is imperative for 'students of antiquity' to defamiliarise it 'because we lose as much in historical understanding as we gain in in convenience' by 'cloaking' the different ancient dating systems with the 'apparently scientific unified weave' of the B.C./A.D. system.[25]

When describing the timeframes of the Greek and Roman world, Feeney primarily emphasises their plurality based on different local traditions:

> In the ancient world, each city had its own calendar and its own way of calibrating past time, usually through lists of local magistrates, just as they had their own currencies, their own weights, and their own religions.[26]

This observation leads straight back to Pétau. What can get lost in speculations about his alleged introduction of absolute time or his relationship to Newton

---

23 Wilcox, *Measure of Times Past*, 7.
24 Feeney, *Caesar's Calendar*, 8.
25 Feeney, *Caesar's Calendar*, 12.
26 Feeney, *Caesar's Calendar*, 10.

and Descartes is that his starting point is the communal or societal character of timekeeping. Wilcox knew this quite well as he cites Pétau to the effect that 'all peoples civilised and barbarian divide up time into civil parts'.[27] It is thus Pétau's fundamental intuition that calendric and chronological systems are coextensive with human society. No society without its system of time measurement. As we deal with a multitude of such communities in past and present, it stands to reason that there must be a diversity of calendric and chronological systems.

Such a starting point, of course, makes perfect sense once we consider what kind of work Pétau wrote, a comparative study of the known calendars with the goal of establishing a universal chronology on that basis. The existence of a plurality of such systems is thus almost inevitably his point of departure. How then did he supposedly come to affirm a rather different kind of time? Is the 'absolute' time he proposed, according to Wilcox and his followers, no longer the product of a specific society? And if so, what does that say about his 'science' of time?

These questions need more careful consideration than they have received to date. They will be discussed in the following chapters based on a new examination of Pétau's own text. Without anticipating the result of this analysis, a likely answer can be surmised at this point. Whatever timeframe or dating system Pétau advocated cannot have been abstracted from his estimation of the society he himself was part of. To the extent, then, that he introduced a more universal dating system, this would reflect his own understanding of the place his own culture occupied in the history of the world. In other words, to the extent that his dating system is absolute, it would point to the ultimacy of Western civilisation within the history of human societies and their various conceptions of time.

Why would Pétau have held such a conception? One rather obvious answer to this question is that it corresponds to his religious conviction that Christianity and in particular Roman Catholicism occupied a unique, universal, and final stage in the history of religion. More recent scholarship has, however, pushed against this tempting assumption. In the wake of Adalbert Klempt's important study, the intuitive notion that the B.C./A.D. dating system reflects Christian universalism has largely been discarded on the grounds that, for Pétau, this reference point was adopted for merely pragmatic reasons ('rein praktische Erwägungen').[28] Some have gone further and claimed that the reason this system was not introduced earlier lay in the impossibility of agreeing on the chronological fixation of the year of the Incarnation.[29]

---

27 Wilcox, *Measure of Times Past*, 205.
28 Klempt, *Säkularisierung*, 86.
29 Feeney, *Caesar's Calendar*, 8.

Be this, however, as it may, there is no doubt that for Pétau the science of time has a religious dimension. He observes that calendars and systems of chronology were generally a matter of priests, and this allows him to draw a line from Babylon and its astronomer priests to Pope Gregory XIII and his recent reform of the Julian calendar.[30] Moreover, in his dedicatory letter to Cardinal Richelieu, he extols the dignity of time by comparing it to God in its omnipresence, eternity, and incomprehensibility.[31] If this is more than rhetorical flourish, it would indicate that knowledge of the true God is instrumental or indeed requisite for gaining a proper understanding of time and, ultimately, for its perfect ordering.

1.3 Pétau and early modern historiography

Pétau's concern with scientific chronology led him directly to the problem of historiography. Wilcox opined that 'fundamental to Petavius' method was a sharp division between chronology and history', but this claim is misleading.[32] Pétau's *De doctrina* most obviously went beyond earlier works in scientific chronology by including a chronicle of world history from the creation of the world until A.D. 533.[33] In the later, popularising *Rationarium temporum*, this aspect is even more prominent – a full chronicle to Petavius' own day takes up the bulk of the work, from Book II to Book X, after a summary of the principles of chronology on the merely eighty pages of Book I.[34]

In one sense, this is unsurprising. Petavius' 'social time' inevitably has two dimensions: on the one hand, it is cyclical and as such concerned with the regular feast days and, more generally, the establishment of a workable calendar. On the other hand, time is linear stretching from the present into past and future. These two dimensions are in view where Pétau divides his *doctrina temporum* into computistics and chronology.[35] The latter of these disciplines arguably has to prove its worth by integrating the multifarious and variegated information contained in the historical sources the early modern erudite has at his disposal. As we shall

---

30 Petavius, *De doctrina*, vol. 1, *Epistola*, sig. a4r: [...] *regendi gubernandique temporis munus apud omnes populos soli quondam sacerdotes, ac pontifices obierint.*
31 Petavius, *De doctrina*, vol. 1, *Epistola*, sig. a3r: [...] *excellentem divini numinis in plerisque similitudinem gerit.*
32 Wilcox, *Measure of Times Past*, 204.
33 This is Book XIII of the work. About its origin see below in Section 3.2.
34 Petavius, *Rationarium*.
35 Petavius, *De doctrina*, vol. 1, *Prolegomena* 3.

see later, Pétau is quite conscious that this kind of competence is the touch stone, so to speak, of his scientific chronology.³⁶

Yet the level of detail to which he takes his own work on a universal chronicle clearly goes beyond what is strictly needed in the context of a work establishing the principles of scientific chronology and betrays Pétau's own interest in the study and writing of history. This interest connects him with one of the most remarkable pursuits among early modern scholars, the debate about the nature of history and the practice of historiography. This debate spawned its own genre, the *Ars historica*, learned works written in the humanist style drawing on classical examples to establish the rules of good historiography. Recent scholarship, above all the work of Anthony Grafton, has done much to elucidate the principles, the arguments, and the obstinate problems characteristic of this discourse.³⁷

Was history a story, or was it fundamentally memory? Did it have to be written by eyewitnesses, and if so, was *historia* actually about the past or not, rather, the present? But if it was about the past, how could the veracity of our sources be established? Moreover, what was the relevance of the presentation of history? Should it be a 'good story' and to this end be embellished, for example, by speeches elucidating the character of main actors, or should it avoid those elements which smacked too much of later fabrication?³⁸

In all its breadth and diversity, however, the debate about the *ars historica* had one shared objective expressed in the title of these books: to establish history as a discipline, an *ars* in its own right. This purpose indicates the link between the fundamentally humanist concern for good historiography with the philosophical problem of the place of history within the Aristotelian system of disciplines which, *nilly willy*, formed the basis of systematic approaches to human knowledge of the time.³⁹ This system was fundamentally based on the distinction between *scientia* and *ars* which ultimately went back to Aristotle but took its decisive shape during the Middle Ages.⁴⁰ Sciences generally dealt with what was necessary and

---

36 In fact, he uses this expression in the title of an apology of his work against a French critic: Petavius (Pétav), *La pierre de touche*.
37 Grafton, *What Was History?*; Seifert, *Cognitio historica*, 12–35.
38 Vossius, *Ars historica*, XX–XXI, p. 96–111. Grafton, *What Was History?*, 34–49.
39 Seifert, *Cognitio historica*, 22.
40 A popular starting point was Aristotle, *Nicomachean Ethics* VI 4, but the first book of *Posterior Analytics* was also important. For the various shapes the distinction of sciences and arts took during the Middle Ages an early modernity see: Roelli, *Latin*. For an early modern account of this distinction see: Zabarella, *De natura logicae* I 2, p. 2: *duo [...] disciplinarum genera, quorum unum in iis rebus versatur, quae a nobis fieri possunt: alterum in iis, quae non a nobis fiunt, sed vel semper sunt vel certas alias causas extra nostram voluntatem positas consequuntur.* The former are the arts, the latter are the sciences as he explains in the remainder of the section.

could be demonstrated: history then could not be a science. Yet it was neither among the group of *artes* that were recognised and taught in the West since late antiquity. Instead, it was seen as belonging to one of the latter, traditionally grammar but in early modernity more typically rhetoric.[41]

Neither of these systematic arrangements seemed too satisfactory, however, at least not from the perspective of the passionate historians of the sixteenth century. This gave rise to the quest for an account of principles and methods specific to historiography and thus capable of establishing it as a discipline in its own right. Yet the origins of this attempt in the language and the categories of Aristotelianism meant that the debate about the *ars historica* could never quite escape from the conceptual constraints inherent in this philosophical system and, therefore, remained exposed to fundamental criticism in the name of the philosophical tradition.[42]

This is where Pétau enters the fray. From the outset of his work, he flags up his philosophical and especially Aristotelian credentials.[43] While certainly not an orthodox Aristotelian – in fact, as we shall see, his use of philosophical terminology is often casual – it seems clear that he seeks to distance himself from the humanist milieu in which the *ars historica* debate takes place. Accordingly, he sides with those who flatly deny the possibility of an art of history arguing that *historia* is pure narration without a genuine concern for causes or explanations.[44]

As such, Pétau urges, *historia* becomes the material of chronology which, by imposing order and structure, turns history into a proper discipline for which, interestingly, he prefers the title *scientia* (while using *ars* occasionally).[45] This is the upshot of Pétau's integration of history into his *doctrina temporum*, the science of time. On its own, *historia* cannot aspire to any disciplinary dignity, but when subjected to the principles of chronology, it takes on the rigorous status of a science.

---

41 Seifert, *Cognitio historica*, 16.
42 See Seifert, *Cognitio historica*, 21. The most emphatic verdict was passed by Zabarella (*De natura logicae* II 24, p. 67) who denied that that history could be *ars* in the Aristotelian sense.
43 Petavius, *De doctrina*, vol. 1, Prolegomena, Προθεωρία, sig. e3v: *omisso praefationum more, quae continenti, et perpetua oratione fere constant, philosophos hoc loco potius, maximeque dialecticos, imitabimur, qui propriis disciplinae suae quaestionibus certa de illius natura, genere, subiecto, principiis, ac proprietate capita praefigunt, quae prolegomena et progymnasmata nominantur.* Subsequently, in Prolegomena 1, he introduces a division of material and formal time (see below Section 4.2.2) with reference to *iisdem illis, quos imitari volumus, artificibus.*
44 Petavius, *De doctrina*, vol. 2, Prolegomena, a2v; *Rationarium*, Ad lectorem, sig. a7r: *Habet hoc historia proprium, uti plenius rei gestae modum ordinemque perscribat: nullis fere neque probationibus et argumentis; neque testibus, unde singulorum annorum ratio constet.*
45 Further on his terminological preference in this regard Section 4.4 below.

This gain, however, comes with a heavy price tag, for the only history that can result from this kind of operation is the tabular arrangement of world events as we find it in annals and chronicles. The written history established as a genre since Thucydides, in Pétau's conception becomes little more than grist to the mill of the scientific chronologist. The chronicles included in *De doctrina* and the *Rationarium* are, then, paradigmatic for the kind of historiography facilitated by scientific chronology.[46] More remarkably still, the same principle is applied without much ado to the histories contained in Scripture – both, secular and sacred history are in need of the chronologist's scientific confirmation. Ultimately, science and religion come together in chronology without caveats or compromises. Both serve human societies through their ability to organise and structure a reality that is otherwise chaotic and unwieldy in its infinite diversity.

1.4 The scope of this study

In what follows, these questions will be further pursued. The basis of the investigation is a close reading of Pétau's own works, mostly the two-volume *De doctrina temporum* (1627) and the more popular *Rationarium temporum* (1633). As far as possible within the limits of this study, Pétau's ideas are contextualised within major intellectual developments of his period: scientific chronology, early modern Aristotelianism, and the debate about historiography. Since all these conversations draw heavily on earlier ideas, notably the literature of Greece and Rome which is still accepted as classical, some use of this broader background will also be made. That said, a full account of Pétau's background must remain the task of further research.

While Pétau is not an unknown figure, serious scholarship focused on his understanding of time and history has been in short supply. Some more work has been done on his historical theology, but even this has been mostly in French and German. There is thus no easily accessible resource for the reader of English to find more than basic information about Pétau as a person, his life, and his most important works. It therefore seemed advisable to begin the present account with a section detailing what we know about Pétau.

This will be followed by a summary of his main work on chronology, *De doctrina temporum*, its content and its purpose. Pétau's motivation to research and write on chronology arose from the perceived challenge to Catholic intellectual dominance in the field by the towering contribution of Joseph Scaliger, the leading

---

46 On early modern interest in tabular history in general see Steiner, *Ordnung der Geschichte*.

European scholar of the generation before Pétau and a proud Protestant. As will be seen, Pétau's relationship to Scaliger's work is ambivalent throughout: while the Jesuit most obviously sought to rebut and criticise Scaliger's scholarship, Pétau in fact followed key insights pioneered by his predecessor thus ultimately continuing his work rather than interrupting it.

The two main chapters are devoted to Pétau's understanding of time and history, respectively. These are obviously not entirely separate topics. Moreover, the nature of his work inevitably gave a kind of hierarchy to the two topics as, after all, history is dealt with *within* Pétau's account of time. As indicated in the Introduction, however, Pétau's account of social time and his intervention in the debate about the nature of historiography respond to different contemporaneous discourses. It therefore makes sense to treat them separately without losing sight of their inner connection.

Two final chapters broaden the horizon by addressing the role of religion in Pétau's conception of time and history before sketching out some connections with his theology, specifically his doctrine of God's eternity.

2 Denis Pétau: A scholar between tradition and modernity

While there is little doubt about Pétau's importance as a seventeenth-century religious scholar and author, he remains little known beyond a small circle of specialists. In fact, no detailed research on Pétau has been published in English in over a century.[1] Who then was Denis Pétau or Dionysius Petavius as he called himself according to the conventions of his time?

To readers such as Donald Wilcox, Pétau has appeared as one of his age's modernisers, but his personal life and career do not mark him out as a radical by the standards of his period. Rather, there is something conventional, even boring, about his biography despite the evidence of intellectual brilliance, hard work, and high achievement to which his career and his works speak. It is nevertheless worth laying out the known facts about his life and his literary production before turning to the question of his scholarly and intellectual character.

2.1 Life

The sources on which any reconstruction of Pétau's biography depend have been fully studied by Michael Hofmann.[2] His succinct account of Pétau's life is therefore definitive and probably as accurate as the historical record permits.[3] Hofmann's principal source is the extensive biographical essay which François Oudin published in Jean-Pierre Niceron's *Mémoirs pour servir à l'histoire des hommes illustres dans la république des lettres* in 1737.[4] As far as possible, Hofmann identified and consulted Oudin's sources.[5] In addition, he relied on letters to and from Petavius as well as an obituary published by Henri de Valois in 1653.[6] Moreover, in an appendix to his Ph.D. dissertation, Hofmann included an unpublished *curriculum*

---

[1] There is a little noted (only partially published) PhD dissertation by Broker, *Influence of Bull and Petavius*; and Sabrey, *The Person and Work*.
[2] Hofmann, *Theologie*, 1 and the two Appendices: 'Anhang I: Zur Biographie Petau's', *Theologie*, 256–257 and 'Anhang II: Bibliographische Notiz', *Theologie*, 258–286.
[3] Hofmann, *Theologie*, 1–4.
[4] Oudin, 'Denis Petau'. He also made use of earlier scholarly works, notably Stanonik, *Dionysius Petavius*. This remains the fullest study of Pétau's life and works. Cf. Stanonik, *Dionysius Petavius*, 7–8 for a description of the biographical material Oudin used for his essay.
[5] Hofmann, *Theologie*, 1.
[6] Pétau's letters are edited in three 'books' in *De doctrina temporum*, vol. 3/ii (Antwerp: Gallet, 1702), 298–362. De Valois, *Oratio*.

Open Access. © 2026 the author(s), published by De Gruyter. This work is licensed under the Creative Commons Attribution-NonCommercial-NoDerivatives 4.0 International License.
https://doi.org/10.1515/978-3-11-2223345-003

vitae compiled by Pietro di Rosa from the Roman archive of the Jesuit order.[7] The following account will therefore rely on the results of Hofmann's research but also draw on some of the earlier sources, notably Oudin's rich and detailed account.

Pétau was born in Orléans on 21 August 1583.[8] His father, Jerôme, was a merchant with humanistic interests; his mother's name was Françoise née Hanapier.[9] One of Jerôme's uncles was Paul Pétau, himself a significant scholar, publisher, and book collector who was *conseiller* of Parliament from 1588 to 1614.[10] On Jerôme, Oudin remarks that he 'applied himself more' to the letters than to his business and as a consequence left little money but an excellent education to his nine surviving children who were all (boys and girls) brought up with the classical languages and wrote poetry in Latin and Greek.[11]

Earlier in the sixteenth century, Orléans was one of the strongholds of French Protestantism, and Jerôme Pétau was apparently himself toying with the idea of converting.[12] Eventually, he doubled down on his Catholicism, and his children were brought up on solid counter-reformation principles. Jerôme soon recognised the unique intellectual gifts of his second son, Denis. Oudin reports that the father conveyed to his son early on the need to vanquish in intellectual battle the man he referred to as *Géant des Allophyles*, Joseph Justus Scaliger (1540–1609).[13]

The story may or may not be historically reliable, but it illustrates well the renown Scaliger enjoyed as one of the most erudite people of his age as well as the extent to which the French Catholic elite felt challenged by his proudly exhibited Protestantism. This did not change in the decades to come, and the need for a critique of his ideas became one of Denis Pétau's main biographical fixtures –

---

7 Hofmann, *Theologie*, 256–257.
8 Hofmann, *Theologie*, 2 and 256; Oudin, 'Denis Petau', 81.
9 Hofmann, *Theologie*, 2. Tellingly perhaps, the mother is not mentioned by either Oudin or Stanonik. Hofmann's diligence discovered it in a genealogical article by one Dr. Fouchon, 'La société de Heere, 23.
10 On Paul Pétau see now the webpage of the Paul Petau Digital project at the University of Heidelberg, online at https://digi.ub.uni-heidelberg.de/en/petau/index.html. Accessed on 10 May 2025.
11 Oudin, 'Denis Petau', 82. There is some disagreement in the sources regarding the number of Jerôme's offspring. Fouchon lists the names of nine boys and three girls of whom three boys died 'en bas âge'. Hofmann, *Theologie*, 2 is surely right to give credence to this information over the slightly divergent count offered by Oudin, 'Denis Petau', 82: 'six garçons et deux filles'. Fouchon clearly used the civic archives of his home city.
12 See Holt, *French Wars of Religion*, 52–55. On the religious life in Orléans during the latter part of the sixteenth century see now Foa, 'Le repaire et la bergerie'.
13 Oudin, 'Denis Petau', 83.

whether this was indeed inspired by filial piety or merely part of Denis' own self-understanding as a Catholic scholar of his own time.

From the solid foundations laid in his parental home, Pétau went on to further studies in philosophy, first for a year at the Collège d'Orléans, then in Paris at the Sorbonne.[14] He defended his theses for his master's degree in Greek and subsequently spent two years attending lectures in theology by such luminaries as André Duval (1564–1638), Philippe de Gamache (1568–1625), and Nicolas Ysambert (1565/9–1642).[15] During his years in Paris, he frequented the Royal Library where he befriended the celebrated Isaac Casaubon (1559–1614). The letters they exchanged were written in classical Greek.[16] Their friendship came to an end only when Casaubon returned to England in 1610 dashing the hopes of his French Catholic friends for his eventual conversion to Catholicism.[17]

Casaubon encouraged Pétau to work on a new edition of the works of Synesius of Cyrene which Pétau eventually did although at first he only gave the public a taster of his art through the Latin translation of Synesius' brief treatise on Dio Chrysostom contained in the full edition of this author's works which was prepared and published in 1604 by the celebrated humanist, Fédéric Morel II

---

14 Oudin, 84. Neither Oudin nor Hofmann attempt to assign dates to these studies, but if Pétau was nineteen years old when he graduated from the Sorbonne (Oudin, 'Denis Petau', 85), the following chronology would seem plausible: 1598 – Study of Philosophy at Orléans; 1599 – Study of Philosophy at the Sorbonne; 1600–1602 – Study of Theology. This is assuming Pétau needed only one year to complete his master's degree but would still imply that he began his studies at age fifteen and moved to Paris as a sixteen-year-old!

15 On these three scholastic theologians see now: Alemanno, *Aspetti della cultura teologica*. Brief biographical sketches on Duval (or Du Val): 47–49; Gamache: 51–52; Ysambert: 57–58. On Duval see further: Calvet, 'Un confesseur de Saint Vincent de Paul'.

16 Pétau's letters to Casaubon are lost (Stanonik, *Dionysius Petavius*, 15) except for one draft in the Archives françaises de la Compagnie de Jésus (Chantilly) which Hofmann has seen (*Theologie*, 270 [B.I.8.c]), but five of Casaubon's letters to Pétau are published as nos. 1028, 1034, 1038, 1044, and 1105 in Isaac Casaubon, *Epistulae*, 598; 600–601; 602–603; 608; 637.

17 According to Oudin ('Denis Petau', 88–89), Pétau from this point considered Casaubon 'ne [...] que comme un ennemi de l'Église, & le réfuta, quand il le trouva en son chemin, mais sans trop le chercher'. Stanonik comes to a more nuanced conclusion: 'Solange Casaubonus lebte, hütete er sich sorgfältig, ihm auf dem Kampfplatze zu begegnen, ja er citierte gelegentlich mit ehrender Anerkennung seine philologischen Arbeiten'. Only years after his death, he found it impossible to avoid 'einzelne seiner Irrthümer zu widerlegen, ohne jedoch die Gelegenheit hierfür zu suchen' (*Dionysius Petavius*, 23). Cf. also Stanonik, *Dionysius Petavius*, 23 n. 66 with relevant citations exemplifying these criticisms. Casaubon's confessional stance was a matter of much speculation. See the brief summary in Considine, 'Isaac Casaubon (1559–1614)'. For a much fuller account: Pattison, *Isaac Casaubon*, 124–127; 144–146; 187–194; 212–216.

(1552–1630), another acquaintance of Pétau's from his early years in Paris.[18] By the time Morel's edition appeared, Pétau had taken up his first position as Professor of Philosophy at the University of Bourges. At the time of his appointment, Pétau was nineteen years old. He stayed at Bourges for two years before returning to Paris in 1604 to complete his theological education.[19]

Early in his career, Pétau decided he would join the ranks of the clergy; he became subdeacon, then canon at Orléans Cathedral.[20] His contact with the Jesuit order, however, did not happen until his time in Bourges where, according to Oudin, he was attracted to them by the opposition they aroused in Protestants and notably in Scaliger.[21] After his return to Paris, it was Fronton du Duc (1558–1624), himself a Jesuit, whose example inspired Pétau to join the order. He undertook his noviciate in Nancy in 1605, completed his theological studies at Pont-à-Mousson and was ordained to the priesthood in 1609. He took his final vows in Paris in 1618.[22]

From 1609, Pétau taught at Jesuit institutions. His first appointment was as Lecturer of Rhetorics at the Jesuit College in Reims where he stayed for three years before moving to the famous Collège Henry IV at La Flèche in 1612.[23] There he remained until 1617, and during this time his research was mostly dedicated to textual editions. In 1612, his edition of Synesius' writings eventually appeared in print, followed by editions of works by Emperor Julian 'the Apostate' (1614) and the Constantinopolitan orator, Themistius (1618).[24]

Pétau's next steps seem somewhat unclear from our sources. He worked on the first-ever edition of the *Short History* (*Breviarium historicum*) by Nicephorus I, the early-ninth century Patriarch of Constantinople, and in connection with this project seems to have spent parts of 1615 and 1616 in Paris.[25] The edition appeared in 1616.[26] Yet at Easter 1617, we find him again at La Flèche now as Professor of Holy Scripture.[27] In 1618, the Jesuits were finally permitted by King Louis XIII to

---

18 Morel (ed.), *Dio Chrysostom*, sig. c2v-f6r. On Morel see now Kecskeméti, *Fédéric Morel II*. On his relationship with Pétau see Hofmann, *Theologie*, 292, n. 9.
19 Oudin, 'Denis Petau', 85.
20 Oudin, 'Denis Petau', 86.
21 Oudin, 'Denis Petau', 86–87.
22 Hofmann, *Theologie*, 2–3.
23 Hofmann, *Theologie*, 3 with n. 20 (p. 294) points out slightly divergent information in our sources about those years. Full account of these years in Stanonik, *Dionysius Petavius*, 20–29.
24 Petavius (ed. and trans.), *Synesii opera; Iuliani orationes iii; Themistii orationes xix*.
25 Oudin, 'Denis Petau', 91. See also the detailed argument in Hofmann, *Theologie*, 295, n. 21–22 concerning Pétau's whereabouts during these months.
26 Petavius (ed. and trans.), *Nicephori breviarium historicum*.
27 Oudin, 'Denis Petau', 91–92.

open their own College in Paris, the Collège de Clairmont. Pétau must soon have been moved there and initially continued to teach Rhetoric until, in 1621, he was appointed to a Chair in Positive Theology which at the time meant something more like historical theology.[28]

In the same year, he brought out the works of Epiphanius of Salamis together with a new Latin translation.[29] Pétau's main theoretical interests come to the fore as never before in the edition of this important fourth-century Christian author, as he adds often lengthy notes on questions of chronology (notably on the years of Christ's birth and his passion), on problems of Church history and the interpretation of key documents from the Patristic era.[30]

These became the central topics to occupy Pétau during the remaining three decades of his life which he spent mostly in Paris as attempts to win him for positions in Spain (1629) and Rome (1638/39) floundered.[31] Probably in 1623, he became the College's librarian, a role he retained until the end of his life, while he gave up his chair in 1644 to focus on his research.[32] Pétau's physical constitution was poor throughout his life. By 1651, things had become so bad he moved temporarily back to Orléans in the hope of some improvements to his failing health. When these expectations did not materialise, he returned to Paris where he died on 11 December 1652.[33]

2.2 Works

Pétau's literary output is extensive. The fullest list is contained in Carlos Sommervogel's *Bibliothèque de la Compagnie de Jésus* with corrections included in an appendix of Hofmann's study.[34] The present account has no ambition to be comprehensive. Pétau's poetic works – tragedies and poems in Latin, Greek, and Hebrew, a Greek paraphrase of the Book of Psalms, and a large number of orations – will

---

[28] Oudin, 'Denis Petau', 93–95. For the idea of 'théologie positive' see Quantin, *Le catholicisme classique*, 103–111 and now Levitin, *Kingdom of Darkness*, 123–138.
[29] Petavius (ed. and trans.), *Epiphanii opera omnia*. According to Stanonik (*Dionysius Petavius*, 34), the work went into print in late 1621 and appeared in early 1622.
[30] Petavius, *Animadversiones*.
[31] Oudin, 'Denis Petau', 113–115 (Madrid); 134–136 (Rome). Cf. Hofmann, *Theologie*, 297–298, n. 30–31 with further references to sources.
[32] Oudin, 'Denis Petau', 157–158. For the chronological discrepancies in the sources see Hofmann, *Theologie*, 297, n. 26.
[33] Oudin, 'Denis Petau', 170–171.
[34] Sommervogel, 'Denis Petau'. This list should now be read in conjunction with Hofmann's corrections and additions: *Theologie*, 258–286 ('Anhang II: Bibliographische Notizen').

be left to one side. So will his various works of theological polemic, mostly against Calvinists and Jansenists.

Apart from those, his early professional years were largely dedicated to editions, the most important of which have already been mentioned. The historical focus of his work was on Greek literature of the fourth and fifth centuries. From this period, he included pagan authors, Julian and Themistius, as much as Christian ones, notably Epiphanius and Synesius.[35] Nicephorus' *Short History* was the major text from the Byzantine period Petavius edited. In many cases, he included Latin translations of his own with the edition of the Greek original. Some of these editions continued to be reprinted and used until the nineteenth century. Pétau's profound knowledge of the classical languages stood him in good stead and often allowed him to propose plausible readings and emendations where his manuscripts were faulty. That said, his access to manuscripts was relatively limited which vitiated the value of his editions compared to those of later scholars, such as the Maurists.[36]

In the 1620s and into the 1630's, Pétau turned his attention to scientific chronology. 1627 saw the publication of Pétau's main work in this field, published in two massive volumes of almost 1,000 pages each under the title *De doctrina temporum*.[37] This work, which stands at the centre of the present study, will be discussed in more detail in a later section.[38] Key insights of Pétau's chronological studies were subsequently included in his *Rationarium temporum*, a single-volume work published in 1633.[39] The first eighty pages of this book summarise the chronological theories from *De doctrina*, whereas the remaining nearly 900 pages contain a full chronicle of world events from its creation to Pétau's own time. The *Rationarium* arguably became his most influential work and was translated into French and English; it remained in use until the nineteenth century.[40] For this success, it is fair to say, Pétau's chronicle was responsible more than his calendric calculations which were moved to the end of the work in later editions. In fact, the English translation of the *Rationarium*, published in London in 1659

---

35 See above p. 17–18.
36 Hofmann, *Theologie*, 9–10.
37 Petavius, *De doctrina*.
38 See below Chapter 3.2.
39 Petavius, *Rationarium*. The work saw a second (revised) edition in 1634, then another five during Pétau's lifetime, and countless more until the nineteenth century. Sommervogel, 'Denis Petau', 599–603 (no. 30); Hofmann, *Theologie*, 263 ('zu Nr. 30').
40 Cf. Ideler's judgment: 'Sein *Rationarium temporum*, welches öfters gedruckt ist, [...] gibt die Resultate seiner chronologischen Untersuchungen in Form eines Handbuchs der Geschichte, das lange das beste in diesem Fach gewesen ist, und wegen des Chronologischen noch immer verglichen zu werden verdient': *Handbuch*, vol. 2, 605.

under the title *A History of the World or An Account of Time*, omitted the summary of Pétau's chronological theories altogether and started straightaway with Pétau's 'chronology' of the creation of the world.[41] Pétau also published *Tabulae chronologicae*[42] and, to defend his chronological theories against an obscure French aristocrat and author, named Jacques d'Auzoles Lapeyre (1571–1641), wrote one of his few French publications, *La pierre de touche chronologique*.[43]

During the final twenty years of his life, Pétau's research was focussed on theology. His most ambitious and undoubtedly most important work from this period shows his abiding interest in time and history. *De theologicis dogmatibus* is an attempt to present the entire contents of Christian doctrine in historical form.[44] While the work is organised around doctrinal topoi (God, Trinity, Incarnation), the treatment of each of these doctrines is geared towards their historical emergence and development. This is specifically significant for the doctrines that are unique to Christianity, such as the Trinity and Christology. The parts of his work dealing with these theological ideas pioneer what later will become known as the history of dogma.[45] Pétau's work rests on a huge compilation of relevant extracts from earlier Christian writers, mostly but not exclusively, from late antiquity. He organises this material, comments on its chronology, changing subtleties in the church's teaching, and potential external influences on doctrinal developments.

Pétau only completed part of the project, but even so its expanse is awe-inspiring its four parts being published in five volumes of just under 1,000 pages each. Theological historians well into the nineteenth century relied on Pétau if only as a

---

41 Petavius, *History of the World*.
42 Petavius, *Tabulae chronologicae*. Sommervogel ('Denis Petau', 597 [no. 26]) lists six full editions until 1706.
43 Petavius (Pétav), *La pierre de touche*. Sommervogel, 'Denis Petau', 604 (no. 32) also lists the various works by d'Auzoles Lapeyre.
44 Petavius, *De theologicis dogmatibus*. The most comprehensive study of the work to date is Hofmann, *Theologie*. See also Leo Karrer, *Historisch-positive Methode*. By far the most scholarly attention has been focused on the single question of whether Pétau believed that the Council of Nicaea in 325 corrected systemic Platonic tendencies among the earlier Christian authors. See Galtier, 'Petau et la preface'.
45 In the nineteenth century, F. C. Baur opined in his lectures on the History of Dogma: 'Es ist daher eine merkwürdige Erscheinung, dass während doch der Protestantismus das wahre historische Bewusstsein und Interesse weckte, gerade die katholische Kirche das erste große Werk hervorbrachte, in welchem die Entwicklungsgeschichte des Dogmas zum Gegenstand einer eigenen umfassenden Untersuchung gemacht ist. Es ist dies das in der Geschichte unserer Wissenschaft Epoche machende berühmte Werk des französischen Jesuiten Denys Petau (Dionysius Petavius) *de theologicis dogmatibus*. ' (*Dogmengeschichte*, vol. 1, 112).

collector of relevant extracts from the primary sources.[46] More intriguingly, from the time of its first publication, the work was devoured by religious dissenters in France and abroad, as it gave them all the ammunition they needed for their argument that the doctrines of the church could not be traced back to apostolic times. It was no coincidence that *On Theological Doctrines* was first reprinted in Amsterdam by the Calvinist, Jean Le Clerc (1657–1736), in 1700.[47]

Among the more traditionally minded, Pétau's work was seen with considerable suspicion.[48] In England, Bishop George Bull (1634–1710) wrote *Defensio Fidei Nicaenae* in 1685 partly in repudiation of Pétau's historical observations.[49] In a remarkable proto-ecumenical gesture, this work was warmly embraced by Jacques-Bénigne Bossuet (1627–1704), the Catholic Bishop of Meaux who, according to Bull's biographer, assured his Anglican colleague of 'the unfeigned Congratulations of the whole Clergy of France'.[50]

This somewhat unexpected reception of Pétau's work has given rise to questions about his own stance on theological and even ecclesiastical matters.[51] This is not the place to pursue those questions, but they conveniently return us to the point from where the present section started: what was Pétau's position in the intellectual controversies of his time? Was he a moderniser and if so, in what sense?

There certainly is no evidence whatever to suggest that he strayed outside the bounds defined for him by his order. He clearly had no sympathies for those who were at odds with the Catholic Church as represented by the Society of Jesus.

---

46 E.g. Dorner, *Entwicklungsgeschichte*, 57, note. Note that this reference is omitted in later editions of the work. Baur tellingly complains in the introduction to his *Versöhnungslehre* that so little work has been done on the history of this doctrine that 'das bekannte Werk des Denys Petau [...] hier nicht einmal den Werth einer reichen Materialien-Sammlung [hat]': *Lehre von der Versöhnung*, 18.

47 Dionysius Petavius, *De theologicis dogmatibus* (1700). On Le Clerc see: Levitin, *Ancient Wisdom*, 90–95. More radical than Le Clerc was Daniel Zwickel, a committed anti-trinitarian who relied heavily on Pétau in his *Irenicum Irenicorum* (1658): Levitin, *Ancient Wisdom*, 483–484. Isaac Newton, too, read and relied on Pétau: Mandelbrote, 'Than This Nothing Can Be Plainer'.

48 See Levitin, *Ancient Wisdom*, 502–523.

49 Bull, *Defensio*. For the historical background see Nelson, *Life of Dr. George Bull*, 280–293; Quantin, *Church of England*, 343–349.

50 Nelson, *Life of Dr. George Bull*, 385.

51 Bull himself cited and rejected the view that Pétau was himself 'secretly' an anti-trinitarian and instead argued that the Jesuit *Pontificiae potius quam Arianae causae consultum voluisse* by implying (a) that the earliest fathers lost their authoritative status and (b) that Ecumenical Councils could establish new doctrines (*Defensio*, 9). Chadwick blamed Pétau's 'crude, unsubtle, oversimplified' obsession with 'the poisonous influence of Platonism' on Christianity for his censure of the early Christian writers (*From Bossuet to Newman*, 58–59. Quantin wisely counsels that 'what Petau had been really aiming at is debatable' (*Church of England*, 346, n. 122).

Pétau was no irenic – he terminated his friendship with Isaac Casaubon when the latter, after his return to England in 1610, made it clear he had no intention of becoming Roman Catholic. In general, he only entertained friendly relations with Protestants where he was hopeful of converting them.⁵² As we shall see in more detail later, much of his chronological work was undertaken to advance the long-standing desire to have a potent Catholic reply (or rather, a riposte) to Scaliger's epochal scholarship, and while technically this was not a matter of confessional polemics, it certainly participated in the wider sense of competition between the religious camps.

The assessment of Pétau as a moderniser as proposed by Wilcox, Klempt and others, therefore needs to be qualified, but this does not make it wrong. As we shall see, there is a breathtaking spirit of renewal and innovation at work throughout Petavius' scholarship. He is fully convinced of progress in history and sees his own world at the forefront of this development.⁵³ Moreover, as Michael Hofmann has argued, he embraces a scholarly ethos of exactitude and precision which leads him to critical assessments of all historical sources including religious texts.⁵⁴

None of these tenets and principles, however, seem to have conflicted with his proudly affirmed attachment to Tridentine Catholicism. In fact, they can be seen as its outgrowth. Marc Fumaroli's observations about the spirit of Jesuit Catholicism in seventeenth century France neatly apply to Pétau:

> The Society of Jesus considered itself contemporaneous to a time of miracles and wonders in which the vital impulse (*élan vital*) of nature responded to the call of divine grace. In this dialogue, the Jesuits themselves played the role of mediator and 'midwife'. In this enthusiastic view of Christian modernity, the resources of the Catholic tradition – from the Apostles and the Fathers to the medieval Doctors, as well as the wealth of pagan and Christian antiquity – offered themselves less from the angle of reminiscence (*réminiscence*) and more under that of memory, a living memory from which a word emanates that calls into actualisation the potentialities of the present.⁵⁵

---

52 Hofmann, *Theologie*, 17. In addition to Casaubon, Hofmann mentions in this connection Lucas Holstein (Holstenius) and Hugo Grotius (*Theologie*, 18–21).
53 He regularly insists on the *novelty* of the science he proposes: *De doctrina*, vol. 1, *Epistola*, sig. a6r, [...] *nunc primum in absolutae scientiae redacta formam, quasi nova quaedam disciplina prodeat*. Later in the *Prolegomena*. Προθεωρία: *Tum demum novam quandam esse scientiam intelliget, atque ab reliquis, quae hactenus in usu feruntur, proprio quodam iure distinctam* (sig. e3r). On the underlying sensitivity of Jesuit learning see the insightful article by Marc Fumaroli, 'Temps de croissance'.
54 Hofmann, *Theologie*, 65–69.
55 Fumaroli, 'Temps de croissance', 152: 'La Société de Jésus s'estime en effet contemporaine d'un temps fertile en miracles et en merveilles, où l'élan vital de la Nature répond à l'appel de la grâce

Pétau therefore does not display any trace of the anti-modern *ressentiment* that later on was to become so characteristic for a certain type of religious writer. Expressed in more theoretical language, Pétau's views on time and history are embedded in a theological frame – in that sense, they are certainly not secularising on their own terms. While he insists on the significance and indeed the normativity of science, there is no whiff detectable that this emphasis could collide with the validity of his religious and ecclesiastical commitments.

It will be the task of the following chapters to show in more detail how Pétau held these different commitments together, what his leading theoretical assumptions were, but also to investigate where faultlines in his theories become visible which could, over time, lead to conceptions and approaches based on remarkably different sets of assumptions. Before addressing the details of Pétau's conception, however, some more general comments are needed on his major work on chronology and its scientific and polemical context.

---

divine, les Jésuites jouant dans ce dialogue un rôle médiateur et "accoucheur". Dans cette vue enthousiaste de la modernité chrétienne, les ressources de la Tradition catholique, depuis les Apôtres et les Pères jusqu'aux Docteurs médiévaux, les richesses de l'Antiquité païenne et chrétienne, s'offraient moins sous l'angle de la réminiscence que sous celui de la mémoire, d'une mémoire vivante d'où jaillit une parole appelant les virtualités du présent à s'actualiser.'

3 *De doctrina temporum* in the context of early modern chronology

3.1 Joseph Scaliger and scientific chronology

Pétau's conception of time and history cannot be discussed in abstraction from his published work on scientific chronology. And his writings on chronology, above all *De doctrina temporum*, cannot be understood without the background of Joseph Scaliger's pioneering contribution to this discipline.[1] Perhaps Pétau would have been drawn to questions of chronology simply based on his own philological and historical work; we cannot exclude that possibility. Even if such a non-polemical path towards his interest in chronology were permitted, however, Scaliger's significance and influence would be undeniable simply because of the way his work, conducted decades before Pétau's own activity, loomed large over a discipline he all but founded.[2]

As it is, Pétau's work on chronology is not merely dependent on Scaliger's scholarship in the way *any* such work at the time would have been, but its very conception is owed to the urgent sense existing in the Catholic world and especially the Jesuit order at the beginning of the seventeenth century that a full rebuttal of Scaliger's work was needed in order to counter the impression that a Protestant could have produced an unrivalled scholarly masterpiece. This purpose is evident on nearly every page of *De doctrina* whose very structure often seems determined by Pétau's complaints about smaller or larger inaccuracies in Scaliger's work.

It is widely accepted that much of that criticism was valid. Often cited is the assessment given (allegedly) by Petavius' Protestant contemporary, Gerrit Vos (Gerardus Vossius):

> Those who want to compare, without emotion or partisan passion, what they wrote about time will learn that, in places, greater praise is owed to Scaliger, while also discovering places where they will much rather agree with Petavius.[3]

---

[1] For the fullest presentation of Scaliger see Grafton's two-volume intellectual biography: *Joseph Scaliger*.

[2] Building on the work of Grafton and others, Dmitri Levitin has made the stronger case for a vast '*programmatic* influence' Scaliger exerted on subsequent generations of scholars, as the 'historical work of the *Thesaurus* [served] as a model to be emulated, almost as a test of any budding scholar's capacity.' 'From Sacred History', 1124.

[3] *Qui sine affectu ac partium studio conferre volet, quae de temporibus scripsere, conspiciet esse ubi Scaligero maior laus debeatur, comperiet quoque ubi longe Petavio malit assentire.* The

Comments such as this one imply that Pétau's *De doctrina* can and should be read alongside Scaliger's books because, despite the polemical presentation the Jesuit author adopts, his book is far from a 'demolition job'. Rather, he ultimately takes a much more nuanced approach to the work of his celebrated predecessor whom he 'attacked, appropriated, and improved', as Anthony Grafton observed.[4] Scaliger *and* Petavius thus became shorthand for scientific chronology and were referenced as such in authors such as Edward Gibbon and Giambattista Vico.[5] In the nineteenth century, Ludwig Ideler, in his *Handbuch der mathematischen und technischen Chronologie*, calls them *die beiden Heroen dieses Faches*, i. e. of scientific chronology.[6]

For the present investigation, Pétau's debt to Scaliger is much more important than are his criticisms. The latter are mostly directed at details of calendric, astronomical, and chronological theory – to questions, that is, which lie outside the scope of the present investigation.[7] By contrast, Pétau's conception of scientific chronology and his methodological approach follow quite naturally in the wake of Scaliger's work even where he deviates from his predecessor's conclusions – and these aspects of his scholarship are key for the present investigation.

---

statement is regularly cited, most recently by Di Rosa, 'Denis Petau e la cronologia', 12 and (in English) by Wilcox, *Measure of Times Past*, 208–209. I have, however, been unable to trace it to any of Vossius' works or his letters. The probable source of all later citations is Oudin, 'Denis Petau', 111, where no precise reference is given. Stanonik, *Dionysius Petavius*, 58, n. 170 clearly takes it from Oudin. Di Rosa offers a more precise reference, '*De historicis graecis*, Praef', but this seems incorrect. Wilcox (*Measure of Times Past* and n. 19 on p. 282) claims Grafton's authority ('Scaliger and Historical Chronology', 175–176) for a reference to Vossius' '*De theologia gentili et physiologica Christiana* ... 3rd ed. (Frankfurt, 1675), 28:212' but this again leads nowhere.
4 Grafton, 'Scaliger and Historical Chronology', 174.
5 Gibbons famously wrote about his early fascination with chronology that 'in my childish balance I presumed to weigh the systems of Scaliger and Petavius, of Marsham and Newton which I could seldom study in the originals; the Dynasties of Assyria and Egypt were my top and cricket-ball: and my sleep has been disturbed by the difficulty of reconciling the Septuagint with the Hebrew computation.' *Memoirs*, 43. Vico, who was fundamentally critical of their work, refers to them as 'i due meravigliosi ingegni, con la loro stupenda erudizione' *La scienza nuova* II 10.2, p. 523. English text: Bergin and Fisch (trans.) *The New Science*, 284.
6 Ideler, *Handbuch*, vol. 2, 603. Di Rosa's ('Denis Petau e la cronologia', 49) conclusion seems appropriate: 'In definitiva, nella fondazione della scienza cronologica, Scaligero, l'iniziatore, aveva bisogno dell'opera perfezionatrice di Petau; come Petau aveva bisogno di quella creatrice di Scaligero. L'opera di Petau è considerata appunto come un complemento e un perfezionamente dell'opera scaligeriana; e – ironia della sorte! – il nome di Petau rievoca sempre quello dell'avversario e viceversa, cosi che il binomio Scaligero – Petau è legato indissolubilmente alla fondazione della scienza cronologica.'
7 The fullest discussion of Pétau's chronology is Di Rosa, 'Denis Petau e la cronologia'. For some comments see Ideler, *Handbuch*, vol. 2, 602–605.

In a seminal article on Scaliger and scientific chronology, Anthony Grafton set out to determine the novelty of Scaliger's work.[8] He quickly dismissed the frequent claim that Scaliger turned chronology into a science by combining for the first time philological with astronomical analysis and including Ancient Near Eastern sources in addition to Greek and Roman ones. Both principles, Grafton explains, were established and widely practised decades prior to Scaliger. What stood out was the utter scholarly quality, the methodological rigour, and the unprecedented scope of Scaliger's scholarship. By applying his stupendous knowledge of sources and his prodigious skills as a critical editor, he reinvented his discipline in practice if perhaps less in theory than has sometimes been asserted.[9]

In addition to the breadth and sophistication of his research, Scaliger's ambition was also novel.[10] Chronology had previously been thought of as ancillary to historiography.[11] Treatises in chronology were, therefore, written mostly for students or other beginners. Scaliger sought to raise the stakes of the debate and initiated a form of self-consciously scientific chronology aimed squarely at the erudites of his age. As we shall see, Pétau agreed with all these novel departures Scaliger had introduced.

Grafton has drawn attention to some letters suggesting that Scaliger came to chronology through an attempt to edit a 'computus', a medieval treatise on the calendar.[12] He apparently worked on this edition from the late 1570s but ultimately extended this task into a much bigger project within which this originally planned edition came to be included, at least in parts. This was his *De emendatione temporum*, published in 1583.[13]

By the time he wrote *De emendatione*, Scaliger had already earned Europe-wide reputation as an editor of classical texts. Among those was the *Astronomica*, a poem in five books by a first-century Roman author, Marcus Manilius, which Scaliger published in 1579.[14] In the process of preparing the edition, Scaliger un-

---

[8] Grafton, 'Scaliger and Historical Chronology', 158.
[9] Grafton, 'Scaliger and Historical Chronology', 158–161; See also Grafton, *Joseph Scaliger*, vol. 2, 25–89 on chronological work done prior to Scaliger. See also Levitin, 'From Sacred History'.
[10] Grafton, 'Scaliger and Historical Chronology', 161–162.
[11] See further on this in Section 5.1. below.
[12] Grafton, 'Scaliger and Historical Chronology', 157–158.
[13] Scaliger, *De emendatione*. Grafton, *Scaliger*, vol. 2, 89–93.
[14] Scaliger, *In Manilii Astronomicon*. Grafton, *Scaliger*, vol. 1, ch. 7. Grafton is at pains to push back against what he calls 'the standard judgement of Bernays and, following him, Pattison', according to which Scaliger *mainly* edited Manilius out of interest in ancient astronomy (*Scaliger*, vol. 1, 186; cf. Bernays, *Scaliger*, 47). Yet Grafton's detailed account of Scaliger's *Commentarius* shows the *particula veri* of Bernays' thesis by observing that 'most of Scaliger's notes treated

dertook a careful study of ancient astronomy which quite naturally paved the way towards his subsequent work on chronology.

De emendatione temporum consists of eight books. In the first four of them, the author investigates as comprehensively as possible all calendars known in the West at the time, encompassing a total of over fifty different systems.[15] In each case, he seeks to establish their astronomical basis and illustrates them through handy tables which were often simply taken over by Pétau and others.[16] In Books V and VI, he investigates all known 'epochs' or eras, by which he means events or points in time from which calendars have been dated, beginning with the creation of the world.[17] On this basis, the dates of events given in different calendric and chronological systems can systematically be compared and synchronised. After that, in Book VII, Scaliger offers a selection of medieval calendric texts – these are the computuses one of which apparently lay at the origin of the whole work's genesis.[18] Finally, Book VIII addresses calendar reform which was highly topical especially after the introduction of the Gregorian reforms in 1582.[19]

Scaliger's results were often critical, undermining long-held assumptions about the synchronicity of events on different chronological timelines. One famous example of his criticism is the popular identification of the Babylonian King Nabonassar (747–734 BC) with King Salmanassar who, according to 2 Kings 18, 9 conquered Samaria. This identification had first been proposed by Copernicus whom many others followed.[20] But, as Scaliger triumphantly pointed out, the former was a Babylonian, the latter an Assyrian king.[21] This was no trifling matter because the accession of Nabonassar, which according to Babylonian records co-

---

astronomical or astrological points' giving 'his reader clear elementary treatments of technical problems' (*Scaliger*, vol. 1, 194).

15 Scaliger, *De emendatione* I–IV.
16 Grafton, 'Scaliger and Historical Chronology', 160. See e.g. Scaliger, *De emendatione* I, p. 19 and Petavius, *De doctrina* I, vol. 1, 61; Scaliger, *De emendatione* II, p. 63–64 and Petavius, *De doctrina* II, vol. 1, 129–130.
17 Scaliger, *De emendatione*, V–VI.
18 Scaliger, *De emendatione* VII.
19 Scaliger, *De emendatione* VIII.
20 Copernicus, *De revolutionibus* III 11, sig. t4v: *Huius rei supremum scopum constituit Ptolemaeus principium regni Nabonassarii Caldeorum, quod apud historiographos in Salmanassar Caldeorum regem cadit.* See Grafton, *Scaliger*, vol. 2, 124–130 for a full discussion of Copernicus' theory and its reception.
21 Scaliger, *De emendatione* V, p. 214: *Qui Salmanassarum eundem cum Nabonassaro faciunt, non advertunt eum Salmanassarum non Babylonis, sed Ninivae regem fuisse. Nabonassar vero Bablyonis rex fuit.* Cf. Grafton, *Scaliger*, vol. 2, 302–303.

incided with a lunar eclipse, had been crucial to chronological calculations by Ptolemy, the greatest astronomer of the Hellenistic age, and a biblical reference to this king would have facilitated the synchronisation of the biblical chronology with that of the Babylonians and thus, with Greek and Roman history as well.[22]

Scaliger was not, however, content with the kind of relative synchronisation made possible by the comparative study of these calendars. Rather, he argued for the need to introduce a single, universal timescale against which all other calendars and the events dated according to them, could be measured. This universal timeline had to be wholly abstract, an invented time, so to speak. Scaliger devised such a timeline and called it the Julian period:

> The creation of things, and other ancient periods of time, both historical and mythical, cannot be marked out exactly unless we invent (*fingamus*) for ourselves a continuous timeline as a norm against which we can arrange all chronological data.[23]

As its name suggests, the Julian period is based on the solar year of the Julian calendar used in the Christian world until the Gregorian reform of the late sixteenth century. To arrive at a suitable cycle, Scaliger adapted an approach first taken by Dionysius Exiguus in the sixth century.[24] Dionysius, in order to compute a reliable Easter calendar, had multiplied the 28-year solar cycle and the 19-year lunar cycle. The resulting cycle of 532 years seemed sufficient for his ecclesiastical purpose. But for Scaliger's more ambitious project, it was too short, so he also included the 15-year Indiction cycle which was widely used in the Byzantine Empire. The resulting cycle has a total length of 28*19*15 = 7,980 years. Within this cycle, each year had its unique character (say S15, L3, I10). The beginning of the cycle (S1, L1, I1) was, according to Scaliger's calculation the year 4,713 B.C. and thus prior to every known historical event.[25]

In fact, it was prior to the year in which, according to the Bible, the world was created. In proposing a 'year' prior to the creation of the cosmos as the starting point of his chronological system, Scaliger did not intend to cast doubt on the accuracy of the Scriptural narrative, but to imagine a timeline against which competing calculations of all world events could be compared. This included the cre-

---

[22] Ptolemy, *Syntaxis mathematica* III 7, ed. Heiberg, 254, 8–13. See also Toomer (trans.), *Ptolemy's Almagest*, 9–12.
[23] Scaliger, *Thesaurus temporum, Isagogici canones*, p. 180: *Conditus enim rerum et alia temporum intervalla vetustissima tam historica, quam mythica, perfecte notari non possunt, nisi nobis continuam seriem temporum fingamus, ad quam, tamquam ad normam, omnes temporum titulos dirigamus.*
[24] See Mosshammer, *Easter Computus*.
[25] Grafton, 'Scaliger and Historical Chronology', 181–185.

ation itself whose date was controversial since the chronologies of the Hebrew Bible and its Greek version, the Septuagint, were at variance with each other.[26]

While these innovations to the study of chronology vastly enhanced Scaliger's reputation as a giant of erudition among his contemporaries, his work was by no means uncontroversial. Scaliger keenly embraced daring hypotheses that seemed intuitively plausible to him even when they were not based on very solid evidence.[27] This led to many skirmishes with colleagues and rivals about various theoretical and scientific matters ranging from seemingly marginal questions, such as the linguistic origin of the name 'Essenes' given to a Jewish sect in Philo of Alexandria,[28] to his idiosyncratic interpretation of the Attic year which was obviously central to his scientific project and which, as we shall see, Petavius took pleasure in rebutting.[29]

Particularly sensitive were cases in which Scaliger walked a fine line between critical historical scholarship and the outright rejection of traditional assumptions. For the adjudication of his sources, Scaliger adopted the methodological principle that the most trustworthy ones are those closest to the event in question. This allowed him to recognise as spurious the widely accepted *Antiquities of Annius*, a fifteenth-century forgery pretending to contain the remnants of historical writings by a range of authors from across the ancient world. Information from Annius' *Antiquities* had seeped into many historical and chronological works throughout the sixteenth and into the seventeenth century with disastrous consequences.[30] Scaliger was not the first to recognise the fraudulent nature of Annius' work, but he certainly stated it with aplomb at the outset of *De emendatione*.[31] Yet willingness to believe these fake sources, for Scaliger, was no worse than the corresponding failure to identify the kernels of accurate historical information contained in the actual sources we possess. To use the language of a later generation of scholars, he insisted on the combination of negative and positive criticism. Ul-

---

26 On the exegetical problem see Larsson, 'Chronology of the Pentateuch'.
27 See the useful review essay by Reiss, 'Early Modern Separation'.
28 This debate involved the Hebraist Johannes van den Driesche [Drusius] (1550–1616) and the Jesuit, Nicolaus Serarius (1555–1609). Scaliger wrote *Elenchus trihaeresii Nicolai Serarii* (1605) in defence of Drusius. See Grafton, *Scaliger*, vol. 2, 507–512.
29 The Attic year was the subject of Book I of *De emendatione*. Grafton offers a detailed analysis in *Scaliger*, vol. 2, chapter 2.1 and 2.2. He summarises: 'For all its errors, forced readings, and misinterpretations, [Scaliger's treatment of the Greek calendar] opened up a richer body of texts and a wider range of questions to apply to them than any previous work or group of works.' (*Scaliger*, vol. 2, 177).
30 Annius, *Commentaria*. On Annius and his forgeries see Grafton, *Defenders of the Text*, ch. 3; Mori, *Historical Truth*, ch. 5.
31 Scaliger, *De emendatione* I, p. 2–3.

timately, it was the latter principle that led to some of the most hostile responses to his great work as he was accused of privileging 'lying pagans over sacred oracles'.³²

The most dramatic and the most consequential instance of this opposition arose in the field of Egyptian history. In the course of his editorial work on Eusebius of Caesarea's *Chronicon*, which was the basis of his *opus magnum*, *Thesaurus temporum* (1606), Scaliger became aware of the earlier Christian chronicle written by Julius Africanus at the turn of the third century.³³ This text in turn contained citations from a Hellenistic Egyptian author, Manetho, specifically a fragment listing thirty-one dynasties of Egyptian kings.³⁴ Scaliger decided that this list deserved historical credence as written by an Egyptian priest, even though it went against the earlier testimony of Herodotus.³⁵

The problem was that, according to the fragment contained in Julius' chronicle, these dynasties lasted for over 5,000 years before they came to an end with the conquest of Egypt by Alexander the Great in 331 B.C. In terms of the biblical chronology, however, this would make them antedate the flood and even the creation of the world.³⁶ Now, Egypt obviously could not have had kings before the world had come into existence, so it appears that either Manetho's dynasties or the biblical chronology had to be wrong. Characteristically, Scaliger did not choose either of those alternatives but instead postulated a separate timeline, which he called proleptic time or 'the first Julian period of proleptic time', into which the duration of Manetho's dynasties could fit without somehow contradicting the truth of the biblical account.³⁷

Scaliger's solution evidently is a halfway house that begs the crucial question of what these years could be if they were not included in the normal history of the world. More intriguing is the question of whether his proposal contains an indirect hint that the chronology of Scripture itself might not be as reliable as people thought? Scaliger certainly opposed those who claimed that the Bible alone was

---

32 Schlichter, *Mythology*, 103. See also Grafton, *Scaliger*, vol. 2, 424–425.
33 Julius Africanus, *Fragmenta* ed. Wallraff. At the outset of his preface, the editor refers to Scaliger's *Thesaurus* as 'the work [that] drew scholars' attention to the author of the first Christian chronicle, Iulius Africanus' (v).
34 Julius Africanus, *Chronographiae* F 43, ed. Wallraff, 94–96; F 46, ed. Wallraff, 100–119. On Manetho see Verbrugghe and Wickershan (eds.), *Berossos and Manetho*, 95–120.
35 Scaliger, *Thesaurus temporum. Isagogici canones*, 317. See the full discussion in Grafton, *Scaliger*, vol. 2, 711–720.
36 Grafton, *Scaliger*, vol. 2, 716.
37 Scaliger, *Thesaurus temporum. Isagogici canones*, 123: *Postulatur ut praesumatur una Periodus Iuliana. Vocetur* prior Periodus Iuliana temporis proleptici: *vel* Periodus Iuliana postulatitia. (Italics in the original).

sufficient to resolve even its own timeline.[38] For example (to cite a celebrated exegetical crux), the detailed chronological data given at the outset of the Book of Ezekiel ('In the thirtieth year, in the fourth month, on the fifth day of the month, as I was among the exiles by the River Chebar, the heavens were opened, and I saw visions of God.') does not, on its own, permit dating this event.[39] Scaliger argued that the biblical prophet took for granted the calendar of the Babylonians among whom his nation lived in exile and therefore referred to the 'thirtieth year' in the era of King Nabopollassar.[40]

Scaliger was thus clear-eyed about the limitations of the Bible as a source for precise chronology *even if* it was otherwise taken as the infallible word of God. Did he go further than that, at least implicitly? This is as hard to know as Pétau's attitude to analogous problems. It is, however, remarkable that the latter, despite the polemical nature of his chronological work, did not launch a major attack on Scaliger's willingness to accept the credibility of a source that seemingly called into question the biblical date of the creation, but confined himself to the comment that it made no sense to 'set the origin of any empire before the flood – but far less before the creation of the world as Scaliger does without any justification.'[41] Pétau points out that Eusebius already rejected the information he found in Julius Africanus, and criticises Scaliger for his dismissal of the fourth-century chronologist as a source.[42] His engagement with Scaliger's perhaps most adventurous hypothesis is crisp and dismissive, but remains at the level of scientific chronology.[43]

This observation applies more generally, as will be borne out in the analysis that is to follow. Pétau's polemic against Scaliger is pervasive, acerbic, and often personal. Yet it is directed at the details of his calendric, astronomical, and philological arguments as well as the chronological conclusions he drew from those. By

---

38 Scaliger, *De emendatione* I, p. 2: *Quod si aliquis sacrae historiae peritissimus, hoc est, qui intervalla rerum gestarum nobilissima certissimis ratiociniis ex Mose et reliquis sacris bibliis explorata habeat, nihil tamen ex illis ad certam epocham historiae Graecae aut Romanae referre possit: quodnam adiumentum is ex eiusmodi diligentia adferre potest aut sibi aut studiosis rerum antiquarum?*
39 For a full discussion of the exegetical problem see Kutsch, *Die chronologischen Daten*, 45–54.
40 Scaliger, *De emendatione* V, p. 218: *Haec aera imprimis biblicae historiae studiosis notanda est, cum sit ea, qua utitur Ezekiel Propheta.* A fuller explanation in the third edition (Geneva, 1629), 398, cited in Grafton, 'Scaliger and Historical Chronology', 168.
41 Petavius, *De doctrina* IX 15, p. 36: *Porro, qui chronologiam et verorum temporum historiam profitetur, proplepticos illos et postulatitios annos omittere debet, nec ante deluvium; multo vero minus ante mundi conditum imperii ullius originem inchoare. Quod sine ullo iudicio fecit Scaligerus.*
42 Petavius, *De doctrina* IX 15, p. 36.
43 Pétau included some more pugnacious comments in his *Animadversiones*, 12.

contrast, the Jesuit scholar largely affirms the scientific foundations and the methodological principles laid down by his illustrious predecessor. In this way, his own work continues the project begun by Scaliger despite the veneer of radical rejection in which he covers his presentation.

That said, Pétau had theoretical ideas of his own. While he took from Scaliger the principles of critical scholarship and the foundational ideas of scientific chronology, his understanding of time and history is largely independent of Scaliger's work. It can and must therefore be analysed on its own terms. Before doing so, however, some brief information must be given on the structure and the general contours of Pétau's main chronological work.

3.2 Pétau's *De doctrina temporum*

Pétau's ideas about time are almost exclusively to be found in his works on chronology, and above all his two-volume *De doctrina temporum*. This work was initially published in 1627 following years of preparatory research that was needed to tackle such a vast topic.[44] As Scaliger before him, Pétau used his editorial work, which was the main focus of the decade from 1612 to 1622, to sharpen his understanding of the philological challenges of ancient chronology. More directly ancillary to his later publication was his study of astronomy and his attempts to discover as much as possible about the calendars of other nations and cultures.

It is tempting to see Scaliger and Pétau as the cliché of polar yet somehow complementary opposites, the ying and yang of early modern chronological science.[45] Scaliger was flamboyant, audacious, brilliant, overbearing in his self-confidence, contemptuous of many of his rivals and generally willing to trust his own intuitions. Pétau lacked all the personal and intellectual attributes that made his opponent an intellectual star. Instead, he pursued his scholarly work doggedly, systematically, and often with the help of his fellow-Jesuits. In a letter we possess, he writes to an unnamed Jesuit in Lisbon requesting his help in preparing *De doc-*

---

44 Stanonik dates the beginning of the 'Vorarbeiten für ein ausführliches chronologisches Werk' to 1611 or 1612: *Dionysius Petavius*, 54.
45 For a rather different assessment, see Pattison, *Isaac Casaubon*, 462. According to the Victorian author, 'the Jesuit scholars introduced into philology research the temper of unveracity which had been from of old the literary habit of their church. [...] It may often happen that Scaliger is wrong, and Petavius right. But single-eyed devotion to truth is an intellectual quality, the absence of which is fatal to the value of any investigation. Jesuit learning is a sham learning got up with great ingenuity in imitation of the genuine, in the service of the church.'

trina. His hope is to be connected with Jesuits living in the East who could furnish him with information about calendars and chronologies in use among the Ethiopians and Copts, but also the Chinese, Japanese, and Indians rendered into Latin *quam fidelissime*.[46]

As for his study of astronomy, it has been generally recognised that his astronomical expertise was far superior to that possessed by Scaliger, and many of the improvements Pétau contributed to the discipline are owed to his improved grasp of this science.[47]

With all these preparations in place, Pétau felt ready to take on the celebrated work of the great Protestant scholar. In the *Prolegomena* he is explicit about his goal to tackle the scholarship to be found in Scaliger's *De emendatione* and his *Thesaurus*, specifically its theoretical part, entitled *Isagogici chronologiae canones*. The intellectual battle 'in this whole volume' (*toto hoc in opere certamen*), he writes, is ultimately with Scaliger since he is the one who emphatically claimed to have perfected the science so that 'no part in its completion and perfection remained to be done' (*ad eius absolutionem et perfectionem pars nulla sit reliqua*).[48]

The structure of *De doctrina* bears out this declaration. Pétau follows the time-tested genre of the polemical work which takes the opponents theses one by one, raises queries about them, adduces observations and arguments, and concludes with their dismissal. The order of books often follows those to be found in *De emendatione*. Like Scaliger, Pétau begins with the examination of the Attic year – in fact, his repudiation of Scaliger's somewhat eccentric reconstruction of the Greek calendar has often been seen as one of Pétau's major scholarly triumphs (Ideler calls it 'sonnenklare Widerlegung').[49]

---

[46] Petavius, *Epistula* III 66, in *De doctrina* 1703, vol. 3, 355: *Cum enim opus quoddam De doctrina temporum adversus Josephi Scaligeri commenta De emendatione temporum aliaque id genus, pridem susceperim, opus ad id fore mihi video, cum alio rerum apparatu, tum variarum nationum kalendariis ac computis, Aethiopum praesertim et Coptitarum; tum Orientalium, ut Sinarum, Japonum, Indorum, populorumque ceterorum; quorum de annis magno ille conatu magnas nugas effutiit. Quas ut in reliquis facile praestiti, sic in illis deprehendere ac patefacere sine propriorum cuiusque Kalendariorum adminiculo non possum. Quare peto illud insuper, ut si quid a vestratibus, qui per illas regiones dispersi sunt, curari obtinerique potest, id mihi per litteras significes. Summa autem haec rei est, ut propriam cuiusque nationis, anni formam, ac mensium dierumque descriptionem, necnon et epochas, sive annorum initia, tum intercalationes ac cetera quae eodem spectant, Latine quam fidelissime ab illis perscripta mittantur.*

[47] Ideler, *Handbuch*, vol. 2, 604: 'Dionysius Petavius, der mit gleicher Gelehrsamkeit [sc. as Scaliger] und nicht geringerem Scharfsinn [...] einen ungleich größeren Vorrath astronomischer Kenntnisse verband'.

[48] Petavius, *De doctrina*, Prolegomena 3, sig. i1v–i2r.

[49] Ideler, *Handbuch*, vol. 2, 602.

The back and forth about this issue offers a fascinating window into the unique challenges of scientific chronology as practiced by these two scholars and shall therefore be summarised here to illustrate their work. The conventional assumption had always been that the Attic year followed a lunar calendar.[50] Yet Scaliger disagreed and argued instead for an idiosyncratic system with thirty-day-long months.[51] One of his arguments for this claim was drawn from a passage in Diodorus Siculus' monumental *Library of History* written in the first century B.C.[52] In Book XIII of this work, the author reports an episode from the so-called Sicilian Expedition (415–413 B.C), a disastrous military adventure of the Athenians from the time of the Peloponnesian War.[53] One night, before that expedition had even started, all the city's hermae, stone pillars with a head above a plain lower section, were mutilated. In the ensuing investigation, one man accused Alcibiades, the general, whom he reported to have seen 'enter the house of an alien about the middle of the night on the first day of the new moon (τῇ νουμηνίᾳ)':[54]

> When he was questioned by the Council and asked how he could recognize the faces at night, he replied that he had seen them by the light of the moon.[55]

Based on this report, Scaliger drew the following conclusion:

> How could it have been the νουμηνία [= the first day of the new moon], and the moon was shining in the middle of the night? Hence the [Greek] months were not lunar months.[56]

---

50 Ideler, *Handbuch*, vol. 1, 255–256.
51 On the problem, other early modern approaches, and Scaliger's solution see Grafton, *Scaliger*, vol. 2, 145–167. Grafton concludes that 'Scaliger's work stands out for its intellectual ambition, its richness in detail, its willingness to depart from ancient testimony and modern dogma in order to reconstruct what he clearly saw as a deeply buried ancient truth. But it sought to fill a need that others felt as well, and differed from them chiefly in its resolute idiosyncrasy.'
52 On Diodorus see now Meeus, Sheridan, and Hau (eds.), *Diodorus of Sicily*.
53 For the historical background cf. Kagan, *Peace of Nicias*.
54 Diodorus, *Bibliotheca historica* XIII 2, 4, 2–4: ἔφησεν εἰς οἰκίαν μετοίκου τινὰς ἑωρακέναι τῇ νουμηνίᾳ περὶ μέσας νύκτας εἰσιόντας. English translation: Oldfather, *Diodorus Siculus*, 131. Online at: https://penelope.uchicago.edu/Thayer/E/Roman/Texts/Diodorus_Siculus/13A*.html#1. Accessed on 21 May 2025. Plutarch offers the same story in *Alcibiades* 20.
55 Diodorus, *Bibliotheca historica* XIII 2, 4, 5–7: ἀνακρινόμενος δ' ὑπὸ τῆς βουλῆς, πῶς νυκτὸς οὔσης ἐπεγίνωσκε τὰς ὄψεις, ἔφησε πρὸς τὸ τῆς σελήνης φῶς ἑωρακέναι. English translation: Oldfather, *Diodorus Siculus*, 131.
56 Scaliger, *De emendatione* I, p. 23–24: *Quomodo poterat esse νουμηνία et media nocte luna lucere? Ergo menses non erant lunares.*

The case illustrates well Scaliger's unique scholarly acumen as well as his intellectual brilliance which permitted him to identify the calendric relevance of an anecdote buried deep in a work on history from the Hellenistic period.

It was Pétau, however, who had the last laugh. As it turned out, Scaliger had stopped reading a bit too early and thus missed the main point of the story which, in Diodorus, ends as follows:

> Since, then, the man had convicted himself of lying, no credence was given to his story ...[57]

In other words, the claim that the moonlight permitted him to recognise Alcibiades on the νουμηνία revealed the accuser as a liar because everyone understood that this night was pitch-dark!

One cannot quite blame Petavius for savouring his triumph on this point. He includes this case among the 'abhorrent' arguments adduced by Scaliger which only 'by hallucination could be held to prove' his point.[58] Having provided his readers with the details missing from his opponent's presentation of the facts, he concludes that

> [...] not only is it wrong [to claim] that it can be argued from this passage that one could on the popular νουμηνία see the moon, but it is obvious that the opposite [conclusion] follows.[59]

The same pattern repeats itself over the subsequent books of *De doctrina* which deal in order with the various calendars Scaliger had already parsed. Individual chapters often indicate in their titles a particular claim from Scaliger's scholarship against which Pétau directs his criticism. From Book VI, Pétau's own positive theories come more to the fore. In Book VII, he offers his own introduction to scientific chronology with sections on the solar and lunar cycles as well as the Julian period which he without hesitation took over from Scaliger. Book VIII deals with the celestial motions of sun and moon and explains their eclipses. Books IX to XII contain the counterpart to Scaliger's examination of 'epochs' giving pride of place to Scaliger's Julian period which is here presented again and in greater detail.

---

[57] Diodorus, *Bibliotheca historica* XIII 2, 4, 7–8: οὗτος μὲν οὖν αὐτὸν ἐξελέγξας κατεψευσμένος ἠπιστήθη. English translation: Oldfather, *Diodorus Siculus*, 131.
[58] Petavius, *De doctrina* I 8, p. 18: ... *per hallucinationem ad idem probandum corrogantur.*
[59] Petavius, *De doctrina* I 8, p. 19: *Quare tantum abest ut ex eo loco constitui possit* νουμηνίᾳ *populari lunam potuisse conspici, ut contrarium apertissime sequatur.* On the distinction between a civil and a popular calendar in ancient Greece see Petavius, *De doctrina* I 1.

Despite the inevitable influence Scaliger's *De emendatione* thus exerted on the structure of *De doctrina temporum*, it would be a mistake to reduce Petavius' work to a polemical engagement with Scaliger's scientific insights and claims. Rather, Pétau had his own constructive theories about the subject matter he was dealing with. What is more, these theories, which had little to do with Scaliger, exerted their own influence on shape and structure of Pétau's work. This can be seen from its division into two parts of which the first is contained in Books I to VIII and deals with τὰ τεχνικά, that is, computistics or the science of calendars, whereas the second (Books IX to XII) is devoted to τὰ ἱστορούμενα, i.e. chronology or the science of historical time. This division has no parallel in Scaliger's work but derives from Pétau's own conception of the science of time, as we shall see in detail later.

To understand the composition of Pétau's work, it is therefore necessary to consider both his attempt to investigate, critique, and improve Scaliger's chronological achievements *and* his aim of advancing his own theoretical agenda. Taken together, they reveal the underlying rationale of Pétau's writing, but they also explain some tensions in the structure of *De doctrina* that can make it seem less elegantly composed than Scaliger's work which it sought to improve.

Be this as it may, it is arguably the logic of Pétau's own approach which comes to the fore in Book XIII which stands apart from the rest of the work. Here, Pétau went beyond Scaliger in offering a first stab at a chronicle containing events from the creation of the world to A.D. 533. Apparently, this was a last-minute decision after it proved impossible to include in this place an edition of ancient texts relevant for Pétau's chronological work.[60] This edition eventually appeared as a separate publication in 1630 under the title *Uranologium*.[61] The chronicle was intended to substitute for this abandoned part to ensure the book reached the length required by the publisher. Since the whole work had to be out by the time of the Frankfurt Book Fair of 1627, however, the chronicle remained incomplete (for the same reason, it also contains a rather embarrassing number of errors).[62]

Despite its apparently rather accidental genesis, the transition from chronological theory to historiographical practice was not an ephemeral aspect of Pétau's scholarship but followed from his notion that chronology proper dealt with τὰ ἱστορούμενα. The sequence of (a) the treatment of calendars (computistics); (b) the discussion of historical chronology; and (c) the production of actual historiography based on these insights makes perfect sense on this basis.

---

60 Stanonik, *Dionysius Petavius*, 56.
61 Petavius, *Uranologium*.
62 Stanonik, *Dionysius Petavius*, 56.

This is borne out, furthermore, by Pétau's publication of full *Tabulae chronologicae* in 1628 and above all, from his *Rationarium temporum* which came out in 1633.[63] As briefly described above, this work merely summarised Pétau's chronological theory to proceed directly to a full and extensive chronicle of the world from creation to Petavius' own time. Theory of time, then, and the practice of historiography belong closely together for Pétau. This will become much clearer in the following sections.

To conclude, Pétau's *De doctrina* is largely determined by its polemical purpose which was to take the shine off Scaliger's achievement and cut to size his accomplishments in chronology. This can make the work seem derivative despite Pétau's success in improving many of Scaliger's insights. What I argue here, however, is that there is more to the work than an attempt to correct his predecessor's mistakes. Even a closer look at the structure of the work indicates Pétau's theoretical ambition. To perceive more distinctly this constructive and at the same time highly original side of Pétau's work, we now have to turn to his discussion of the foundations of his theory of time.

---

63 Petavius, *Tabulae chronologicae*. Sommervogel notes that 'ces tables sont imprimées en grandes feuilles' ('Denis Petau', 597, no. 26). Unfortunately, I have been unable to locate a copy. Petavius, *Rationarium*. The *Tabulae* were included in the 1703 *editio novissima* of the *Rationarium* (Paris: Florentinum de Laulne, 1703).

4 Pétau's theory of time

It will be the task of this chapter to consider in detail Pétau's theory of time which underlies his scientific chronology. Pétau's theoretical ambition in *De doctrina* is clear from the outset. While much of the book's content is devoted to rather technical discussions of calendars and their chronological consequences, the work's ultimate aim is well captured by its title: to offer a *doctrina temporum*, a theory or science of time into which these technical insights can be inscribed. This science, as Pétau emphasises again and again, is a novel accomplishment which only became possible in his own era and of whom he is the proud originator.[1]

The foundations of this doctrine are developed early on in *De doctrina*, in the first sections of the Prolegomena to the first volume. The current chapter will therefore to a considerable extent consist of a close analysis of these pages. From this will emerge the contours of a remarkably original and fascinating view of time. Pétau admittedly was hardly a philosopher of the first order. Indeed, the execution of his ideas often lags behind his ambition. In some cases, conclusions are therefore uncertain, even speculative or at least not fully substantiated by his explicit words. This roughness of his account does not, however, make his ideas less interesting or less worth considering. Ultimately, I will argue that Pétau's theory of time is as fascinating as it has been neglected.

Early in the Prolegomena to the first volume of *De doctrina*, Pétau declares that those masters (*artifices*) whom he intends to follow distinguish in the science of time between its material and its formal object. This is one of a number of references to the Aristotelian background of his approach.[2] In his dedicatory letter, he had already announced his intention to pursue his topic 'from causes and principles […] in accordance with the philosophers',[3] and in the initial section (Προθε-

---

[1] See e.g. Petavius, *De doctrina*, Prolegomena, Προθεωρία, sig. e3r.: *Tum demum novam quandam esse scientiam intelliget* [sc. lector], *atque ab reliquis, quae hactenus in usu feruntur, proprio quodam iure distinctam.* On the broader significance of his time see Prolegomena 2, sig. e5v: *Crescentis haec erant, nec ad plenum expolitae artis ac perfectae publicam in rem officia. Cui deinceps aetas nostra, quantum longinquiore intervallo et observatione incrementi et splendoris attulit, tantum ab ea vicissim commoditatis accepit.*

[2] Petavius, *De doctrina*, Prolegomena 1, sig. e3v: *Duplex ab iisdem illis, quos imitari volumus, artificibus disciplinae cuiusvis obiectum, (ita enim nominant) vel subiectum ponitur. Nam alterum infinitum est et, ut materies illa, quae cum forma composita, complexaque naturam perficit, late fusum est, atque aliena circumscriptione terminatum. Unde et* materiale *dicitur. Alterum certa modificatione formaque comprehenditur, quod ideo formale in scholis appellant.*

[3] Petavius, *De doctrina*, Epistula, sig. a5v: *Etenim quicquid illud est, quod contemplationi nostrae praecipua quapiam ratione sic obiicitur, ut ex causis, principiisque cognoscatur, id, ut Philosophis placet, sui generis scientiam efficit.*

ωρία) of the Prolegomena, he explains his decision to replace the *Praefatio* of the rhetorical style with Prolegomena with his intention to 'imitate the philosophers [...], above all the dialecticians, which preface the questions proper to their disciplines with certain chapters on the nature, genus, subject, principles and properties [of their topic]'.[4]

Despite this declared intention, however, Pétau's appropriation of Aristotelian principles is rather casual not to say slapdash. For this, his initial claim that time had to be considered in its material and formal aspect is quite characteristic. Pétau does not say who the 'masters' were who held this view, what this distinction amounted to more specifically, or how it matters for his own approach. Aristotelianism, it seems, offered to Pétau a convenient point of departure along with some useful conceptual tools.[5] He is far removed, however, from the scholastic Aristotelianism still *en vogue* among many philosophers and theologians of the time, not least in his own order.[6] This approach can make it difficult to identify the more specific people or works on which he draws, especially since he does not include references to authors let alone works or passages he may have in mind. It is therefore necessary to begin with the clarification of Pétau's own definition of time and its scientific study before, in a second step, considering its background in philosophical discussions of his time.

---

[4] Petavius, *De doctrina*, Prolegomena, Προθεωρία, sig. e3v: *Philosophos [...] maximeque dialecticos imitabimur, qui propriis disciplinae suae quaestionibus certa de illius natura, genere, subiecto, principiis ac proprietate capita praefigunt*. See also Prolegomena 1, sig. e4r: *scientiae nomen non vulgari aliqua ratione, sed subtili, ac Peripatetica definitione meruerit* ... Note also that he opened the Προθεωρία with a direct reference to *Physics* IV 13 (222b16–19): *Veterum plerique sapientissimum rerum omnium tempus appelarunt: quidam e contrario stupidissimum, indoctissimumque esse dixerunt*.

[5] Cf. in this connection the dismissive comments Scaliger offers at the beginning of *De emendatione* on the difference between his approach and an Aristotelian one: *Nam qui ante nos hanc provinciam aggressi sunt, si modo hanc nostram, non aliam aggressi sunt, ii satis de tempore et eius natura disputarunt. Sed hanc disputationem melius interpres quarti φυσικῆς ἀκροάσεως sibi vindicasset. Neque vero nos id agimus, ut definiamus tempus esse hoc secundum Peripateticos, aut illud secundum Stoicos aut Academicos. Qui istis definitionibus diu immorati sunt et hac sola scientia Chronologiae scribendae modum terminarunt, illi satis verborum quidem, sed rerum nihil definiverunt* (p. 3).

[6] For an incisive overview see the articles collected in Blum, *Early Modern Aristotelianism*. More specifically on the problem of time cf. Daniel, 'Treatments of Time'.

4.1 The definition of time

Having set out with the distinction of time as matter and form, Pétau continues to assert that in its material aspect, time is infinite like matter is more generally. Only in combination with form does it produce the nature of time in its reality. Formally, on the other hand, time is understood in a certain specification and form (*certa modificatione, formaque comprehenditur*).[7] Pétau's fundamental claim, then, is that time is twofold and can be distinguished as being indeterminate and infinite on the one hand, but finite and structured on the other.

From this initial distinction, Pétau immediately proceeds to the observation that time can be considered in various ways and is therefore studied by different sciences. In this connection, he lists the following disciplines:[8]

1. Time can be approached in its nature, in term of its general causes and principles. This is the task of physics.
2. Time can be understood insofar as it is the measurement of the movement from which time is born, that is, the rotation of heavens and the stars. This is the task of astronomy (*astrologia*).[9]
3. Time can be a matter of the brevity and the length of its parts, and its division into different distances between sounds and movements. In this connection, time is studied as part of music.[10]

These three disciplines dealing with time, Pétau writes, are established. And yet, there is another sense of time which is no less deserving of its own foundational and constitutive science (*fundandae constituendaeque scientiae non minus idonea*). This is time insofar as it can be applied to the use and management of human affairs.[11] Its study, Pétau argues, is mandated because there is no known human so-

---

[7] See the text cited above in n. 2 of the present chapter.
[8] Petavius, *De doctrina*, Prolegomena 1, sig. e3v: [*Tempus*] *autem pluribus modis considerare licet. Aut enim naturam illius, generalesque causas et principia; atque ipsam, ut sic dicam, eius essentiam inquirimus, quod totum naturali scientiae, quam physicam vocant, attributum est; aut ut motuum, ex quibus praecipue tempus oritur, mensura est ac regula: coelestium videlicet orbium atque siderum: ita ad astrologiam pertinent; aut ut brevitate ac longitudine partium indicitur et agitationem vel sonorum dispari intervallo temperature: sic ad rhythmicen vel musicam potissimum refertur.*
[9] Pétau uses *astronomia* and *astrologia* interchangeably for the *science* of the heavens. For this tangled use of these terms in pre-modern times see Losev, '"Astronomy" or "Astrology"'.
[10] On music in Renaissance scholarship see Moyer, *Musica Scientia*.
[11] Petavius, *De doctrina*, Prolegomena 1, sig. e3v: *Reliqua est quarta postremaque ratio, atque ut a superioribus diversa, ita fundandae constituendaeque scientiae non minus idonea; cum in tempore id spectatur unum, quatenus ad civilem humani generis usum, tractionemque conformari potest.*

ciety that does not structure its civic life by means of temporal units. Even the most barbarous nations use the solar and lunar cycles to distinguish days, months, and years.¹² As societies become more advanced, they refine and improve, but do not principally replace this approach. Of this practice there can and should be a science, the word taken not in any vague or colloquial sense, but, Pétau emphasises, in the strict sense in which this term is understood by Aristotle and his school (*scientiae nomen non vulgari aliqua ratione, sed subtili, ac Peripatetica definitione meruerit*).¹³

This first sketch Pétau gives of his project is rather vague and leaves many questions open. How are the three disciplines dealing with time related? What does it mean to propose a fourth approach to time in addition to them? Finally, what does any of this have to do with Aristotle or the Aristotelian tradition? Things get somewhat clearer where the author revisits the same issues a few pages further on in Chapter 3 of the Prolegomena. Pétau now clarifies that his fundamental understanding of time conforms to Aristotle's famous definition in *Physics* IV 11 according to which time is 'the number of change with regard to the before and after'.¹⁴ From this definition it follows, Pétau continues, that the study of time as 'number' cannot be separated from the 'science of numbers', nor can it be without reference to the movements of sun and moon.¹⁵

There is the slight complication that in the place of the author's earlier reference to the science of music he now mentions mathematics, for which he does not offer any explanation.¹⁶ More importantly, however, Pétau provides a clue to the

---

12 Petavius, *De doctrina*, Prolegomena 1, sig. e3v: *Etenim nulla tam inculta, tamque barbara gens unquam extitit, quae non partitione aliqua temporis, ac descriptione civilia sua intervalla digesserit. Nam et diurna solis et menstrua lunae spatia et annuas utriusque conversiones cum simplices, tum invicem implicatas et permistas* [sic, sc. *permixtas*] *barbari pariter et humanitate perpoliti homines intellexerunt* [...].
13 Petavius, *De doctrina*, Prolegomena 1, sig. e4r.
14 Aristotle, *Physics* IV 11 (219b1–2): τοῦτο γάρ ἐστιν ὁ χρόνος, ἀριθμὸς κινήσεως κατὰ τὸ πρότερον καὶ ὕστερον. Note, however, that Pétau never refers to the Greek text of the *Physics* or any other Aristotelian writing (at least in his chronological works). This suggests that his approach to Aristotle is fundamentally shaped by the Latin commentary tradition rather than an engagement with his original text.
15 Petavius, *De doctrina*, Prolegomena 3 sig. e6v: *Denique com tempus ex se mensura sit motus secundum priores, posterioresque partes, ut Aristoteli placet, idemque numerus quidam sit earum, quas metitur, partium; non potest ab numerorum scientia illius esse disiuncta cognitio. Itaque non modo sine motuum solis ac lunae, verum etiam sine numerorum et calculorum usu cassus et inanis est illius disciplinae studiosorum labor.*
16 He does say a little later (sig. e6r), but apparently only *exempli gratia*, that *arithmeticae musica, si tamen scientia est diversa, subiicitur, quod illa numerorum uniusmodi ac per sese, haec*

sense in which, in his mind, the three different disciplines that so far investigate time are connected. In the first place, there is physics which provides the definition of time. Insofar as this definition relies on the concepts of 'number' and 'movement' (change), the disciplines dealing with numbers and with celestial movements are inevitably implicated in the study of time.

Pétau's approach now looks less slapdash than it did at first sight. While being anything but a scholastic Aristotelian, he apparently worked on the basis of the Stagirite's definition of time from the outset seeking to inscribe the disciplines traditionally involved with the study of time into the logic of this very definition. Can this insight, then, also illuminate Pétau's central claim that a new approach to time, a novel science, was needed? Can it help us understand how he conceived of the key aspect of his own project?

In order to answer these questions, a first, relevant observation is that Pétau distinguishes between two kinds of time which he calls 'material' and 'formal'. The former he describes as 'vague', 'common', and 'infinite', whereas the latter is structured and made intelligible by the application human calendric systems. As it turns out, elements of this idea can be found in the Aristotelian tradition; Pétau's own conception nevertheless is remarkably original.

4.2 The Aristotelian background

4.2.1 One time or many?

The unity of time was regularly discussed among students of Aristotle in the Middle Ages and early modernity.[17] At one level, this question simply arose from the wording of the Aristotelian definition itself: if time is the 'number of motion', would this not imply that there are as many different times as there are motions? To this, Averroes replied that Aristotle here referred to the principal and most fundamental motion, that of the celestial sphere, the *primum mobile*. As all other cosmic motions are derived from this first one, they all also depend on its time so there is in fact *one* time for all beings in the universe.[18]

---

vero ποιόν τινα et affectum quodam modo spectat; prout disparibus intervallis sonorum concentum efficit.

[17] Cf. Trifogli, 'Unicity of Time'. For the early modern debate see: Bexley, 'Quasi-Absolute Time'.
[18] Averroes, *In physicam* IV, t. c. 132, ed. Venice 1562, f. 203rb F: *Et in quibusdam scripturis invenit s[ecundu]m hoc igitur quod tempus est motus, secundum hoc est numerus unius motuum. Est igitur propter hoc numerus motus continui simpliciter, no[n] numerus alicuius motus. Et intendebat s[ecundu]m haec verba q[uonia]m, s[ecundu]m quod tempus accidit motui et mot[us] est prior*

This view was widely accepted among medieval Christian readers of the Stagirite, notably by Albert the Great and Thomas Aquinas.[19] Some commentators, however, took a somewhat different approach. John of Jandun, for example, contended that, while the time of the *primum mobile* was one and primary, there were also other times based on the plurality of motions across the cosmos, albeit in a secondary sense.[20]

This strictly Aristotelian, exegetical debate subsequently became intertwined with a second issue which had its origin in the Christian theology of creation: the question when exactly time began. Granted that time was created by God, this still left open a number of options between which the decision was not necessarily easy or straightforward. Augustine, for example, pondered in *De civitate dei* XII whether angelic creatures had their own time prior to the existence of the physical cosmos.[21] For Aquinas who, unlike Augustine, accepted the Aristotelian theory of time in principle,[22] the problem arose what kind of time could have existed prior to the creation of the firmament on Day two according to Genesis 1.[23]

---

naturaliter, necesse est ut sit numerus unius motus primo et essentialiter: et est motus qui accipitur in definitione temporis. D[einde] d[icit] [sc. Aristoteles]: *Est igitur propter hoc numerus motus, i[d est] corporis coelestis.* Cf. Trifogli, 'Unicity of Time', 786.

19 Trifogli, 'Unicity of Time', 787. See also: Mansion, 'La théorie aristotélicienne'.

20 John of Jandun, *In Physicam* IV, qu. 28, f. 133b: *Dicerem ergo quod tempus primo et principaliter dictum est unum et idem numero respectu omnium temporalium , sicut motus primus, s[eu] primi mobilis est unus et idem respectu omnium mobilium: et de isto t[em]p[or]e intelligitur quod non possunt esse plura tempora aequalia simul: quorum unum non est pars alterius: ut duo dies et duo menses et huiusmodi. Sed tempus secundario dictum non est unum et idem numero apud omnia: immo multiplicatur s[ecundu]m numerum eorum quae actu moventur & talia bene possunt esse simul plura et aequalia: quotum unum non est pars alterius.* As Bexley, 'Quasi-Absolute Time', 7–9 shows, Suárez held an extreme version of a 'pluralistic' theory according to which there were as many times as there were motions – at least in terms of what he called 'intrinsic time'.

21 Augustine, *De civitate dei* XII 16, 33–46: *usque adeo autem isti [sc. angeli] omni tempore fuerunt, ut etiam ante omnia tempora facti sint; si tamen a caelo coepta sunt tempora, et illi iam erant ante caelum. At si tempus non a caelo, verum et ante caelum fuit; non quidem in horis et diebus et mensibus et annis (nam istae dimensiones temporalium spatiorum, quae usitate ac proprie dicuntur tempora, manifestum est quod a motu siderum coeperint [...]), sed in aliquo mutabili motu, cuius aliud prius, aliud posterius praeterit, eo quod simul esse non possunt; – si ergo ante caelum in angelicis motibus tale aliquid fuit et ideo tempus iam fuit atque angeli, ex quo facti sunt, temporaliter movebantur: etiam sic omni tempore fuerunt, quando quidem cum illis facta sunt tempora.*

22 Cf. Augustine's dismissive comments on the primacy of the movement of the celestial sphere in *Confessiones* XI 23, 29: *si cessarent caeli lumina et moveretur rota figuli, non esset tempus quo metiremur eos gyros et diceremus aut aequalibus morulis agi, aut si alias tardius, alias velocius moveretur, alios magis diuturnos esse, alios minus? aut cum haec diceremus, non et nos in tempore loqueremur aut essent in verbis nostris aliae longae syllabae, aliae breves, nisi quia illae longiore*

In response to this difficulty, medieval theologians developed the concept of the *aevum* as a time above time but within creation.[24] The complex history of this idea cannot be pursued here, but the two strands, the Aristotelian and the more theological ones, appear intermingled and reoriented in a set of Aristotelian commentaries close to Pétau's own time, the so-called Coimbra Commentaries.[25]

These commentaries were produced in the late sixteenth century by Jesuits at the *Collegium Conimbricensis*, the College of Art in the Portuguese city of Coimbra. They encompass eleven books, originally printed in five volumes between 1592 and 1606, and soon became a standard point of reference across Europe. They were frequently reprinted and regularly used especially within Jesuit institutions.[26] It stands to reason that Pétau who taught Aristotelian philosophy for two years at Bourges, knew the Coimbra Commentaries.[27]

The first of these volumes contains the *Physics* commentary, written by Manuel de Góis (1543–1597).[28] Subsequent to the actual commentary on Aristotle's treatise on time in Book IV, the work included among a number of *Quaestiones* a consideration of whether Aristotle's definition of time was appropriate. In its second section, de Góis addressed the idea that time is twofold.[29] In arguing this point, the author starts with the distinction between the primary time of the first-moved and the plurality of times according to the many motions in the cosmos, as previously held by John of Jandun.[30]

---

tempore sonuissent, istae breviore? [...] sunt sidera et luminaria caeli 'in signis et in temporibus et in diebus et in annis'.

23 Thomas Aquinas, *In libros sententiarum* II d. 12, qu. 1, art. 5, respondeo ad 3: *[...] motus caeli incepit secunda die; sed non omnia simul creata sunt; unde non potest intelligi de tempore quod est numerus motus primi mobilis; sed oportet quod vel per tempus significetur aevum, ut quidam dicunt, vel tempus large sumatur pro numero cuiuscumque successionis, ut sic tempus primo creatum dicatur quod mensurat ipsam creationem rerum, qua post non esse, in esse res prodierunt.* Aquinas analogously took Rev 10, 5 to prove that physical time will end in the eschaton. *Quaestiones disputatae. De potentia*, qu. 5, art. 5, respondeo ad 11: *sicut motus caeli deficiet, ita et tempus deficiet, ut per auctoritatem Apocalypsis inductam apparet.*

24 See Porro, *Forme e modelli*.

25 For what follows see De Carvalho, 'The Concept of Time'.

26 Cf. De Carvalho, 'The Concept of Time', 353–357 for a detailed description of the corpus and its various editions and reprints.

27 Oudin comments that 'il lut avec application les anciens philosophes & mathematiciens. On les méprise beaucoup à present, par ce qu'on ne les connoît pas [...]' ('Denis Petau', 85–86).

28 Collegium Conimbricensis, *In Physicam*. On Góis see De Carvalho, 'Manuel de Góis'.

29 Collegium Conimbricensis, *In Physicam* IV 13, qu. 1, art. 2, p. 541–542.

30 Collegium Conimbricensis, *In Physicam* IV 13, qu. 1, art. 2, p. 541: *videlicet bifariam tempus usurpari uno modo prout est numerus, seu mensura motus summe aequabilis, ac primi, qualis est*

In what follows, however, Góis pivots to the more properly theological distinction of physical time (which depends on the movement of the celestial sphere) and a more general kind of time which exists independently of that movement.[31] This leads to a strange inversion of the original division of time as it is now the cosmic time that is more narrowly delimited compared with a truly universal time no longer bound to the physical cosmos.

For this more comprehensive time, Góis does not use the term *aevum* but instead refers to it as 'imagined time' (*tempus imaginarium*). It is thus clear that the distinction he has in mind is no longer primarily responding to the theological problems predominant in the Middle Ages. Rather, Góis aims at a kind of absolute time which is 'more ancient, more even, and more universal' than physical time and, while it cannot be observed in nature, needs to be postulated in order to explain the cosmic, positive time which the Stagirite had in mind in his definition.[32] In calling this time 'imagined', Góis adds, he does not mean it is imaginary but that

> [...] in such a time we apprehend a certain succession connected as if by instances. In this way, all movements, all [moments of] rest and the times themselves are measured.[33]

---

conversio supremae sphaerae: altero quatenus est mensura cuiusque motus et in qualibet re actu motum subeunte insidet. For Jandun's view see in this chapter n. 20 (this chapter) above.

31 Aquinas' claim (cited at n. 23 [this chapter] above) that time will come to an end with the eschatological destruction of the heavenly sphere is *secundum priorem sumitur*, whereas Augustine's 'potters' wheel' (see n. 22 [this chapter] above) is said *secundum posteriorem*.

32 Collegium Conimbricensis, *In Physicam* IV 13, qu. 1, art. 2, p. 542: *Adverte etiam praeter hoc tempus singulare, dari aliud unum quoque in essendo et universale in mensurando, nempe tempus imaginarium, quod illo est antiquius, aequabilius et universalius. Antiquius, quia tempus primi mobilis coepit cum eius motu, tempus vero imaginarium initio caret. Aequabilius, quia reliqua tempora, si spectentur praecise comparatione motuum, quibus adaequantur, alia celerius, alia tardius fluunt, sicut et ipsi motus, paremque extensionem cum illis fortiuntur, ut progressu etiam dicemus; at tempus imaginarium aequa semper labitur velocitate; quia a nullo motu pendet. Universalius, quia non alia tantum, quae sub temporis mensuram cadunt metitur, sed ipsius etiam motus primi mobilis et eius durationis regula est.*

33 Collegium Conimbricensis, *In Physicam* IV 13, qu. 1, art. 2, p. 542: *Verumtamen in eo longe deteriorem conditionem sortitur, quod nequaquam sit aliquid positivum et reale, sed imaginarium: non quasi in sola imaginatione tanquam figmentum consistat; sed quia ut in spatio imaginario, [...] dimensiones quasdam veris dimensionibus respondentes concipimus, ita et in huiusmodi tempore apprehendimus successionem quandam velut instantibus copulatam, quae motus omnes et quietes, ac tempora ipsa metitur.* It is intriguing to compare here Scaliger's use of 'invent' (*fingamus*) for his Julian Period. See Scaliger, *Isagogici chronologiae canones*, 180 and n. 23 in Section 3.1 above. Cf. also Suárez' notion of *successio imaginaria* in *Disputationes metaphysicae* 50 IX and Bexley, 'Quasi-Absolute Time', 10–15. On the medieval background of 'imaginary time' see Edwards, *Time and Soul*, 30–31.

Comparing this conception with Pétau's brief exposition of his own approach shows similarities, but the differences may be more instructive. While both authors distinguish two kinds of time, Pétau's interest clearly is not in contrasting a more restricted, cosmic time with a more universal, imagined time. For the Portuguese author, cosmic time needs to be underwritten by an ideal time that is the condition for the possibility of cosmic time – a time that is transcendental in the later, Kantian sense. For Pétau, by contrast, the problem with cosmic time is not that it is limited, but that it is undifferentiated, 'vague', and 'infinite'. The contrast for him, then, is more that between unstructured and structured time.[34]

4.2.2 Material and formal time

This difference is further illustrated by the prominent use Pétau makes of the duality of time as matter and form which has no counterpart in Góis. Such a distinction was, however, known in medieval and early modern Aristotelianism from its presence in Averroes' *Commentary on Physics*. This text was easily accessible in the seventeenth century because it was included with the Aristotelian text in the so-called Giuntine edition originally published in Venice between 1550 and 1552.[35]

Averroes introduced the distinction of material and formal time in order to address a notorious difficulty in Aristotle's theory of time, namely, its nod to subjective time. Towards the end of his treatise on time, Aristotle raised the problem of

> [...] whether if there were no soul there would be time or not. For if it is impossible for there to be something to do the counting, it is impossible also that anything should be countable, so that it is clear that there will not be number.[36]

---

[34] One might consider also that Pétau has in mind the contrast of time as continuous and time as a series of instances, for which cf. Daniel, 'Treatments of Time', 590–597. However, his reference to mathematics/music in connection with cosmic time (and his use of Aristotle's definition) speaks against this assumption.

[35] The more accessible version, however, is the one published between 1562 and 1574 which has also been reprinted in the 20th century. On the differences between the two editions see: Burnett, 'Revisiting the 1552–1550 and 1562 Aristotle-Averroes Editions'.

[36] Aristotle, *Physics* IV 14 (223a21–24): πότερον δὲ μὴ οὔσης ψυχῆς εἴη ἂν ὁ χρόνος ἢ οὔ, ἀπορήσειεν ἄν τις. ἀδυνάτου γὰρ ὄντος εἶναι τοῦ ἀριθμήσοντος ἀδύνατον καὶ ἀριθμητόν τι εἶναι, ὥστε δῆλον ὅτι οὐδ' ἀριθμός. English translation: Coope, *Time for Aristotle*, 159.

The logic seems simple enough: if time is 'the number of change' according to Aristotle's definition, there has to be someone who counts. No time then without awareness of time. Yet the question of how this ought to be understood turned out to be tricky and answers remained controversial. Consequently, the problem continued to be debated throughout late antiquity, in the Arab world as well as the Western Middle Ages.[37] Often, the hypothesis was considered that the 'soul' without which time cannot exist would be a cosmic soul though Christian authors were generally reluctant to accept this kind of entity.[38]

The medieval discussion was dominated by the contrasting visions of the Arabic philosopher, Ibn Rušd (Averroes), on the one hand, and Thomas Aquinas on the other.[39] Whereas Aquinas (implausibly) suggested that Aristotle had meant the question to be answered in the negative, Averroes understood Aristotle to affirm that time could only exist if there was a soul counting it.[40] Yet this raised the spectre of merely subjective time. To avoid this consequence, Averroes argued that time existed potentially in the cosmic motions but was only actualised in the mind:

> [...] motion would exist, even if the soul did not exist. And insofar as before and after in motion are numbered in potency, time exists in potency, and insofar as they are numbered in act, time exists in act. Therefore, time does not exist in act, unless the soul exists, but it exists in potency, even if the soul does not exist.[41]

---

[37] For the reception in late antiquity see Zachhuber, *Time and Soul*. For a full account until the thirteenth century see: Jeck, *Aristoteles contra Augustinum*. For the early modern debate see Edwards, *Time and Soul*.

[38] Zachhuber, *Time and Soul*, chs. 2–4. For the (limited) reception of the notion of world soul in early Christianity see Zachhuber, 'The World Soul in Early Christian Thought'.

[39] For Aquinas see Snyder, 'Thomas Aquinas and the Reality of Time'. For Averroes: Trifogli, 'Averroes' Doctrine'.

[40] Aquinas, *In Physicam* IV, lect. 23, n. 5 [n. 629]. Averroes, *In Physicam* IV, t. c. 131, ed. Venice 1562, f. 202 rb D-E: *Cum sit declaratum quod, cum numerans non fuerit, no[n] erit numerus et est impossibile aliquid aliud numerare praeter animam et de anima intellectus, manifestum est quod, si anima non fuerit, non erit numerus; et, cum numerus non fuerit, no[n] erit te[m]pus.*

[41] Averroes, *In Physicam* IV, t. c. 131, ed. Venice 1562, f. 202 rb F: *[...] motus erit, etsi anima non erit. et secundum quod prius et posterius sunt in eo numerata in potentia, est tempus in potentiali et s[ecundu]m quod sunt numerata in actu, est tempus in actu. Tempus igitur in actu no[n] erit nisi anima sit; in potentia vera, licet a[n]i[m]a no[n] sit.* English translation: Trifogli, 'Averroes' Doctrine', 60.

According to Aristotle, time was based on cosmic motion but was not identical with it.[42] Averroes therefore suggests that motion existed independently of the soul. Its temporal ordering can thus be said to have extramental existence too, but insofar as time involves counting, it needs to be actualised by a soul. This is because 'outside the mind', according to Averroes, numbers too exist 'in potency' only.[43] Two stones, we might say, exist as such regardless of whether there is someone counting them, but they are not two *qua* two unless someone applies numbers to them.

Ultimately, Averroes concludes that Aristotle's definition of time as the 'number of motion' means that [...]

> [...] the substance of time, which has the role of form in time, is number, and that which has in it the role of matter is continuous motion, since time is not number in an absolute sense, but the number of motion.[44]

Time in its 'material' reality, then, is *motus continuus*, but in its formal perfection it is number, that is, the imposition of structure on the mere homogeneity of material time.[45]

Averroes thus, like Pétau distinguishes time as matter and form. Both moreover contrast cosmic 'time' as mere continuity from formal time as the imposition of numbered structure on its matter. That said, Pétau is not concerned with the problem of subjective time – or is he?

---

42 Cf. Aristotle, *Physics* IV 11 (219a8–10): ὥστε ἤτοι κίνησις ἢ τῆς κινήσεώς τί ἐστιν ὁ χρόνος. ἐπεὶ οὖν οὐ κίνησις, ἀνάγκη τῆς κινήσεώς τι εἶναι αὐτόν. On the interpretation of this notion see Coope, *Time for Aristotle*, 31–43.
43 Averroes, *In Physicam* IV, t. c. 131, ed. Venice 1562, f. 202 rb E: ... *esse numeri in a[n]i[m]a no[n] est omnibus modis esse in a[n]i[m]a, q[uonia]m, si ita esset, esset fictu[m] et falsum, ut chimera et hircocervus; sed esse eius extra mente[m] est in pote[n]tia propter subiectum propriu[m], et esse eius in a[n]i[m]a est in actu, s[cilicet] q[ua]n[do] a[n]i[m]a egerit illam actione[m] in subiecto preparato ad recipiendu[m] illam actione[m] quae d[icitu]r numerus.* English translation: Trifogli, 'Averroes' Doctrine', 59.
44 Averroes, *In Physicam* IV, t. c. 109, ed. Venice 1562, f. 187 ra C: [...] *substa[n]tia te[m]poris quae est in eo quasi forma est numerus, et quod e[st] in eo quasi materia est motus co[n]tinuus, q[uonia]m non est numerus simpliciter, sed numerus motus.* English translation: Trifogli, 'Averroes' Doctrine', 60.
45 On the reception of Averroes' interpretation in early modernity see Edwards, *Time and Soul*, 26–34. Cf. on the distinction of material and formal time in particular Benedictus Pererius, *De communibus omnium rerum*, 386.

4.3 Subjective time as social time

An intriguing possibility arises at this point. Is the subjective dimension of time for the French Jesuit based on societies' concern for time and its measurement? In other words, is his answer to the age-old question of Aristotle's readers, 'Who counts time?', that this is humanity in its social dimension? Pétau is insistent that human communities have *always* relied on some sort of temporal order.[46] Being temporal in this sense, that is, both *concerned* with time and *dependant* on it, they make their own contribution to our understanding of time. This is why, in Pétau's view, the study of human time measurement is not simply a form of history or, as we would say, cultural studies, but part of the science of time and indeed its very core.

It must be admitted that there is something immediately striking about this interpretation. It explains neatly not only Pétau's use of 'material' and 'formal' time, but even more his claim that the social use of time adds a necessary dimension to the study of time as practiced in physics, mathematics, and astronomy. As much as these three disciplines were spun out of Aristotle's definition of time as the 'number of motion', the fourth and final one would be extrapolated from the notion that Aristotelian time is only actualised in the mind. Moreover, whereas the 'subjective' interpretation of Aristotle's theory was inevitably faced with the problem of intersubjective intelligibility and validity, Pétau's 'social time' would offer a surprisingly novel solution to this difficulty.

That said, the parallels between Pétau's theory and Averroes' version of Aristotle's theory of time must not be exaggerated. Averroes is not interested in a duality of times, but in time as existing potentially in cosmic motion and actualised in the soul. This extramental 'matter' is not yet time but only, we might say, the foundation of time. Here, we need to recall Góis and his distinction of cosmic and 'imagined' time. Despite his use of the Aristotelian language of 'matter', Pétau like Góis thinks that cosmic time is 'real'. After all, it is studied by mathematics and astronomy.[47] What societies add is time as structured and determined.

If Pétau, then, does not follow Averroes in thinking of material time as only existing in potency, he does seem to hold that time can only be known properly

---

[46] Petavius, *De doctrina*, *Prolegomena* 1, sig. e3v: see n. 12 (this chapter) above for the full citation.
[47] One might speculate that, according to Pétau, mathematics and astronomy do not after all study 'time' as such but rather phenomena connected with time. Such an interpretation would work well with Pétau's claim that social (i.e. formal) time is the only time we can know and understand (see following note). Yet elsewhere he is perfectly willing to accept that mathematics and astronomy study time: *De doctrina temporum*, *Prolegomena* 1, sig. e3v. Full citation in n. 8 (this chapter) above.

in its 'formal' or social form. As he explains in his dedicatory letter, 'time can be studied by us insofar as it has been adapted to human uses and bent to popular systems of measurement.'[48] Social time, then, is intelligible time. Cosmic time is its infinite nurturing ground but as such remains inscrutable and unknown. As we shall see, this duality of times has far-reaching consequences for Pétau beyond his approach to chronology; it connects his idea about time with his understanding of religion and his theology.

4.4 A new science

The clarification of Pétau's appropriation of Aristotle's definition of time permits the reconstruction of his proposed new discipline in the study of time. Despite his initial suggestion that he merely adds one more perspective to the study of time, his intention is in fact much more radical than that. In proposing to investigate time insofar as it 'could be applied to the civil use and management of the human race',[49] Pétau really aims at the investigation of time in the only way in which it can be intelligible to us. The new discipline will therefore not merely examine one aspect of time, it will study time as it can be approached by human beings.[50] In that sense, the title of his work, *De doctrina temporum*, needs to be taken literally.

Yet this is not a reductionist approach to time as social time. This is clear from the way Pétau goes out of his way to connect his new discipline with the existing ones which, as he previously asserted, deal with time in its material dimension. In his proposed new science of time, there is a division and hierarchy of disciplines. Specifically, there are four disciplines into which this science can be divided. Two of them he classifies as 'higher' on the grounds that they provide knowledge of causes and principles (διότι), whereas the lower ones lead to merely factual

---

48 Petavius, *De doctrina, Epistula*, sig. a5v: *Tempus* [...] *sic a nobis considerari potest, quatenus ad humanos accommodatur usus et <ad> populares rationes inflectitur. Erit ergo peculiaris et ab aliis separata temporum scientia.*
49 Petavius, *De doctrina, Prolegomena* 1, sig. e3v: [...] *ad civilem humani generis usum tractationemque conformari potest.*
50 Petavius, *De doctrina, Epistula* sig. a5v: *Etenim quicquid illud est, quod contemplationi nostrae praecipua quapiam ratione sic obiicitur, ut ex causis, principiisque cognoscatur, id, ut philosophis placet, sui generis scientiam efficit. At est eiusmodi tempus: quod sic a nobis considerari potest, quatenus ad humanos accommodatur usus et <ad> populares rationes inflectitur. Erit ergo peculiaris, et ab aliis separata temporum scientia.*

knowledge (τὸ ὅτι) citing in this connection Aristotle's *Posterior Analytics* I 10 as his authority.[51]

Building on this distinction, Petavius explains that astronomy and mathematics are the two higher disciplines.[52] These are, as we have seen, the two disciplines studying time as 'material subject'. As it turns out, then, the study of time from the perspective of human society is heavily reliant on these sciences as it derives its principles on the one hand from the knowledge of the movement of celestial bodies, on the other hand from the arithmetic calculations of their ratios and divisions (the length of a solar day, the number of such days in a year etc.).

The lower disciplines, by contrast, are computistics and chronology.[53] Of those, the former is defined as ordering time for its social use by formulating canons and rules derived from the higher disciplines. This, in other words, is the science of calendars and in general the order of time by means of which important dates, feasts etc. are determined. Chronology, by contrast, uses certain signs and indications to assign the remembrances of past things to their respective dates. From this, Petavius derives this further distinction that computistics concerns present and future, while chronology deals with the past.[54]

---

51 Petavius, *De doctrina*, Prolegomena 3, sig. e6r: *Hinc illa videlicet nata disciplinarum apud eosdem philosophos divisio, ut aliae superiores et quasi magistrae artes habeantur; aliae inferiores, ac subiectae. Quarum, ut in I Poster[iorum]. Anal[yticorum] cap. x tradit Aristoteles, cum sit idem plerumque nomen, in eo tamen discrimen situm est, quod τὸ ὅτι, hoc est rem ita, ut cognitum est habere se simpliciter nosse sit inferiorum: ad alias vero τὸ διότι pertineat: quae comprehensarum ab istis rerum principia ratiocinando ac demonstrando colligunt, ut istarum demum propria conclusio sit: illarum autem, enunciationes et ἀξιώματα e quibus conclusio deducitur.* Quite why Pétau thinks that the relevant passage is in Chapter 10 is somewhat hard to see. The distinction is certainly introduced by Aristotle at the very beginning of the book (*Analytica posteriora* I 1, [89b21]): ζητοῦμεν δὲ τέτταρα, τὸ ὅτι, τὸ διότι, εἰ ἔστι, τί ἐστιν.
52 Petavius, *De doctrina*, Prolegomena 3, sig. e6v: *[...] scientia temporum totidem superiores habet astrologiam et arithmeticam. Primum conditione ipsa materiae. Tempus enim, circa quod versamur, ut in diurna, menstrua, annuaque spatia, aut longiores etiam orbes includitur, ad astrologiam pertinet, cuius est coelestium corporum , quorum ex conversione tempus oritur, ipsiusque adeo temporis, qui illius est mensura, propria tractatio. Idem vero, quia non continuatum, ac perpetuum, sed in particulas divisum et, ut ita dicam, numerosum, ac modulatum obiicitur, id sibi arithmetica suo quodam iure vendicat. Hinc alterum illud efficitur, ut ab ambabus confecta iam, et constituta principia mutuetur, quae approbare nequeat ipsa per sese.*
53 Petavius, *De doctrina*, Prolegomena 3, sig. i1r: *usum inferioribus relinquit artibus. Quae numero duae ex hac scientia temporis efflorescent. Harum altera computistica vulgo dicitur altera non minus trito, sed commodiore vocabulo chronologia nuncupatur.*
54 Petavius, *De doctrina*, Prolegomena 3, sig. i1r: *Chronologia ars est, quae certis notis et indiciis praeteritarum rerum memoriam suis temporibus assignat. Etenim nulla nisi praeteritis temporis ut historia, sic chronologia potest esse. In quo tam a priore computistica quam ab ea, quae utramque continet, temporum scientia differ. Nam computandi ars non magis ad praeteritum, quam ad*

The 'lower' disciplines clearly deal with time as appropriated and measured by human societies. This appropriation, we might say, happens in two ways, as cyclical and linear. Societies are temporal on the one hand by inscribing their lives into the rhythm of days, weeks, months, and years while, on the other hand, their temporality consists in their memories of the past and their awareness of history. These two modes of temporality correspond to Pétau's distinction between computistics and chronology. Is the former really only about present and future? This seems less plausible than the assertion that chronology proper, the record of time in terms of past events, concerns exclusively the past.

Ultimately, these two disciplines may well be 'lower' in terms of their scientific status, yet they make up the science of time properly. This is clear from Petavius' division of his *De doctrina* whose first part deals with computistics, while the second treats chronology.[55] The full definition of the 'science of time' which Petavius provides at this point makes the same point, 'a science which enquires into the conditions and properties of time so that it may be brought to the use of humanity.'[56]

In view of Pétau's own project, one may wonder why he emphasises the dignity of astronomy and mathematics in the way he does. The answer lies in his conception of science. We have already cited his insistence that the discipline he proposes is science in the proper, Aristotelian sense. A similar protestation can be found at the outset of the Prolegomena where Pétau rejects with vehemence the confusion of his project with what is '*vulgo* called Chronology' (*quae vulgo Chronologia dicitur*). Now, such an assumption could arguably arise from the fact that *scientia* or *doctrina temporum* could appear to be a somewhat loose

---

praesens aut futurum pertinet, et administrandi per sese temporis nudas sine ἀποδείξει regulas praescribit.

55 To complicate matters, Pétau refers to this division as between τὰ τεχνικά and τὰ ἱστορούμενα (*Prolegomena* 7, sig. i5v), but it is clear that the two parts correspond to the two 'lower' disciplines of his science: *Tomi duo sunt. primus τὰ τεχνικά comprehendit, secundus τὰ ἱστορούμενα temporum. In illo disputantur ea, quae ad artificiosum temporis contextum, vel σύστημα pertinent, qualis est illius partium [...] In his scientia ipsa praesertim temporum et ἐπιστημονικὰ illa principia consistunt. Accedit et ex duabus scientiae subiectis artibus prior, quam computisticam nominant, quae ex solo hoc genere principiorum efflorescit. Alter tomus chronologicae arti dicatus, omnia fere, quae in ἐποχῶν et intervallorum ratione controversa sunt, vel obscura complectitur.*

56 Petavius, *De doctrina, Prolegomena* 3, sig. i1r: *Quam sane χρονικήν deinceps nominemus ac definiamus ita si lubet, ut sit scientia, quae temporis, ut ad usum transferri potest hominum, conditiones, ac proprietates inquirit.* There is thus a tension here between Pétau's claim that astronomy and mathematics are the 'higher' disciplines within his *doctrina temporum* and his insistence that computistics and chronology are the two proper parts of this science. One might resolve this tension by speculating that the former do not study time properly but only incidentally. See n. 47 (this chapter) above.

translation of the Greek *chronologia*.⁵⁷ In the wake of Scaliger's scholarship, Pétau is adamant that what he proposes is *scientific* work.

The foundation of this Aristotelian idea of science, according to Pétau, is the notion of demonstration in *Posterior Analytics* I.⁵⁸ He repeatedly emphasises the importance of providing explanatory causes and principles, τὸ διότι, in this connection.⁵⁹ Causal explanation makes a science, and as Pétau is adamant that his own chronology is intended to fulfil this criterion, he insists on the use of disciplines capable of producing this kind of knowledge, namely mathematics and astronomy.

Less obvious is the epistemic status of disciplines that cannot aspire to the rigour of the sciences. Remarkably, Pétau seems mostly uninterested in the concept of *ars* as a specific type of discipline. Aristotelians such as Giacomo Zabarella considered the science-art distinction fundamental for their system of knowledge: sciences produce speculative knowledge, while *artes* are oriented towards practice.⁶⁰ Pétau, by contrast, frequently uses the two terms interchangeably and regularly refers to the same discipline as both *ars* and science.⁶¹

The duality of types of subjects Pétau has in mind comes out nicely where he illustrates it with the rationalist and empiricist schools of medicine in antiquity. Of those, Pétau explains (based on Cornelius Celsus⁶²), the former 'professes knowledge of hidden causes', whereas the latter 'relies solely on use and experience'.⁶³ Rationalist medicine evidently corresponds to his concept of science; dis-

---

57 Note here the somewhat dangerous terminological ambiguity created by Pétau's use of the term chronology. After all, the discipline to which Pétau's whole work contributes is conventionally called 'scientific chronology', whereas he wants to restrict this term to its historical part. His protest against the conventional use of the term at the outset of his work (*Prolegomena*, Προθεωρία, sig. e3r: *Eam [sc. temporum doctrinam] si quis cum illa, quae vulgo chronologia dicitur, confundendam existimet, nae is plurimum opinione falletur.*), may partly be intended to stem this kind of misunderstanding. It nevertheless stands to reason that his terminological idiosyncrasy can easily create confusion in the minds of his readers and students.
58 For the interpretative problems of establishing Aristotle's own view see Barnes, 'Aristotle's Theory'.
59 See n. 51 (this chapter) above.
60 Zabarella, *De natura logicae* I 2, p. 1–4.
61 He notably refers to the 'higher' disciplines as *superiores et quasi magistrae artes*: *De doctrina*, *Prolegomena* 3, sig. e6r. Cf. also *Prolegomena* 2, sig. e4v the reference to *omnes elaboratae, propagataeque artes, tum astrologia* [...]. Occasionally, the two terms are distinguished, see *Prolegomena* 6, sig. i5r: *Quaedam sunt necessaria [...] alia scriptorum fide [...] constant. [...] In illis proprie scientia: in istis ars et prudentia cernitur.*
62 Cornelius Celsus, *De medicina* I, prooemium.
63 Petavius, *De doctrina*, *Prolegomena*, vol. 2, sig. a4r: *Praeclare meo judicio Cornelius Celsus, auctor valde bonus, cum in duas sectas divisam esse medicinam dixisset; unam rationalem, quae*

ciplines that are not sciences, then, are *empirical* and operate based on experience. They provide, according to Pétau, the factual description, τὸ ὅτι. As such, they are indispensable, and he arguably felt that in an ideal system of knowledge, they have their contribution to make as much as the sciences make theirs.

As we shall see later in more detail, Pétau here effectively draws on the specifically early modern idea of *historia* as a type of discipline that relies purely on factual knowledge.[64] As Arno Seifert has shown, this terminological quirk, based on Aristotelian usage, paved the way for the influential modern concept of the empirical.[65] It is in the context of this *Begriffsgeschichte* that the duality of τὸ ὅτι and διότι emerges as a shorthand for two different types of knowledge, knowledge of facts and knowledge of causes.[66] Pétau interestingly uses *historia* in precisely this sense of factual knowledge but only applies it to what we call history. At the same time, the conceptual duality of empirical and causal knowledge forms the conceptual basis of his *doctrina temporum*.

Ultimately, Pétau is less interested in a duality of subjects or disciplines but rather in the mutual contribution of different kinds of knowledge to his own science of time. His concern is not to deny to computistics and chronology the status of a science but to argue that in order to gain it, they need to rely on the findings of astronomy and mathematics. This methodological and practical imperative is expressed, in theoretical terms, through the distinction of τὸ ὅτι and διότι, empirical and causal knowledge. As we shall see in more detail later, these two ideas, that his own work is scientific and that scientific work depends on the knowledge of principles, is characteristic for Pétau's argument throughout.

4.5 Conclusion

In the opening pages of *De doctrina*, Pétau offers a rather fascinating theory of social or cultural time. It is, I have argued, spun out of Aristotle's definition of

---

abditarum et morbos continentium causarum notitiam profitetur: alteram ἐμπειρικήν, quae usum sibi tantum et experientiam reservat mox allatis utriusque partis argumentis, mediam inter diversas sententiam amplectitur et eam medendi rationem anteponit, quae ex utraque composita, ita singularum opportunitates, ac praesidia colligit ut vitia et incommoda declinet.
64 See below, Section 5.2.
65 Seifert, *Cognitio historica*. A recent full investigation of the same topic acknowledged that 'Seifert's work remains the most important contribution to the history of *historia* as an epistemological category ...': Pomata and Siraisi, *Historia*, 5. Cf. Seifert, *Cognitio historica*, 31 where the authors 'confirm Seifert's view of *historia* as the "godmother" of early modern empiricism'.
66 Seifert, *Cognitio historica*, ch. 3.

time in *Physics* IV 11 and thus equates humanity's consciousness of time in general with the need for a subjective dimension of cosmic time. Time, in this sense, depends on human societies (although, as we shall see, these societies relate to time as part of their religious relationship to God). But societies in their turn also depend on time; they are 'temporal' in a strong sense. Pétau believes there is no society without a concern for time. This social form of temporality comes in two forms: it is cyclical thus leading to the establishment of calendars; but it is also linear, namely, as history. Pétau's project is to study social time in these two dimensions. In order to do so scientifically, however, constant attention to the 'higher' disciplines of astronomy and mathematics is required. It is only by following this principle that there can be a truly *scientific* theory of time.

5 Pétau on history

Pétau clearly believed that attentiveness to history was a key part of humanity's social temporality. As we have seen, he argued that societies dealt with time in two dimensions, namely, the calendric structure imposed on present and future, and the chronological structure imposed on humanity's past. His science of time therefore included chronology as one of two main parts. The analysis of the previous chapter has shown in some detail how he intended to guarantee the scientific treatment of this discipline.

Pétau's concern for chronology inevitably raised the question of his understanding of history. In many ways, his high-level description of chronology could almost be mistaken for an account of history. If societies inevitably keep a record of things past; if an awareness of their past is part of their temporal existence: is this not the same as saying that societies have a history of which they retain memories? This suspicion can be furthermore confirmed by the observation that Pétau's title of the chronological part of his work refers to τὰ ἱστορούμενα. That said, Pétau's chronology with its unique emphasis on *dating* past events seems much less than what we would think of as history. Does Pétau, then, think that history can be reduced to the accumulation of chronological information? Or is there a sense of history in which it can be distinguished from chronology? And if the latter, what can such an understanding of history amount to?

As we shall see in what follows, Pétau does indeed introduce a concept of history in *De doctrina* which he differentiates from chronology. Yet the precise way in which he proposes this distinction is rather puzzling, at least at first sight. The first task of the present chapter will therefore be to present and analyse Pétau's own exposition of history in its relation to chronology. In a second step, this will be contextualised against contemporaneous discussions about history and chronology. From this clarification, Pétau's methodology of historical chronology will be elucidated in the third section. This leads directly to his emphatic endorsement of the B.C./A.D. dating system, whose import will be considered in the fourth section.

5.1 History as distinct from chronology

It might initially appear that history does not feature in Pétau's science of time. Petavius' presentation of his 'new' science, which I have followed in the previous chapter, hardly mentions history or if it does, merely refers to it in an aside.

This only changes in Pétau's brief introduction to volume two of *De doctrina*. Given that Pétau explained on the title page that this volume would 'deal with *temporum τὰ ἱστορούμενα*' which could be loosely translated as 'the historical account of the ages', some reference to history would indeed appear pertinent. Yet even in this place, Pétau does not dwell on the matter in much depth. He remains mostly concerned about the distinction between his 'scientific' chronology (= science of time) and chronology in the vulgar sense of ordering events of human history.[1]

By contrast, the author declares that he considers it 'less awful' (*minus verendum*) were someone to confuse chronology and history. Both, he explains, share their relation to the superior science of time. They have the same status and matter, even though they have different names. Both are based on the recollection of events in the past (*rerum [...] gestarum ac praeteritarum memoria*).[2] There is ultimately one difference between them:

> History provides the matter of things to the other one. In return, [chronology] supplies [history] with order and arrangement which is so to speak, the form.[3]

Without the 'protection' of chronology, therefore, Pétau writes 'history is maimed, defenceless and blind'.[4] In support of this assessment, he cites the saying according to which 'chronology is the eye of history'.[5]

---

1 Petavius, *De doctrina*, vol. 2, Prolegomena, sig. a2r-a2v.: *Quamobrem nemo hanc fere temporum disciplinam aliam ob causam expetit, quam ut chronologiam certius, faciliusque condiscat. A quo studiosorum hominum instituto et consilio uti non abhorremus, sic errori quorundam obsistimus, qui cum magistra temporum et principe scientia Chronologiam re ipsa et appellatione confundunt. Sunt enim ambo haec a se invicem genere, materia, principiis, definitone distincta. appellato quidem ipsa fere eadem utriusque. Quod et aliis in disciplinis caeterisque rebus accidit, in quibus universi generis nomen a precipua quadam vel notissima parte, formave transfertur.*
2 Petavius, *De doctrina*, vol. 2, Prolegomena, sig. a2v: *Minus illud verendum arbitror, ne quis chronologiam ab historia non secernat. Quanquam ad eam propius, quam ad superiorem illam temporum scientiam accedit. Nam ab illa, conditione ac materia non differt: quamvis appellatio sit dispar amborum. Utraque enim rerum in gestarum, ac praeteritarum memoria consistit: tempusque non aliter, quam ut elapsum est, respicit.*
3 Petavius, *De doctrina*, vol. 2, Prolegomena, sig. a2v: *Hoc tamen inter illas interest quod historia rerum silvam ac materiam praebet alteri. Haec illi vicissim ordinem et collocationem, quae est veluti forma, suppeditat.*
4 Petavius, *De doctrina*, vol. 2, Prolegomena, sig. a2v: *Huius orbata praesidiis manca, et inermis, ac caeca est historia.*
5 Petavius, *De doctrina*, vol. 2, Prolegomena, sig. a2v: *Itaque tritum illud, quia verissimum erat, in proverbium abiit: Historiae oculum esse Chronologiam.*

The same line of thought is picked up again in Pétau's later *Rationarium temporum*, a popularised version of *De doctrina*. Right at the outset of this work, he defines the two terms as follows:

> History is narration of the past. Chronology, by contrast, is the alignment of this [narration] to its times and the parts of times from certain indications. Therefore, [history] provides the matter and the abundance of detail to [chronology] whereas [chronology] applies the form and as it were the rule to [history][6]

The reference to history as the 'matter' of chronology is recurrent in both texts indicating Pétau's considered view on this point. In the *Rationarium*, however, the author goes on to provide some more detail of how he understands the two terms. About history, specifically, he writes in his prefatory *Ad candidum eruditumque lectorem* that it presents only 'the measure [*modum*] and order of events without any proofs, arguments, or witness accounts (*testibus*), by which the account of each year may be proved correct'.[7]

While these various comments are never too detailed, some picture emerges, but it is a puzzling one. History is nothing but narration, the reporting of events without any attempt to solidify or evaluate them. It is the raw material consisting of bits of information we can glean from our sources prior to any critical attempt to sift or filter them. If we take the Aristotelian language of 'matter' and 'form' seriously, 'history' is shapeless and chaotic until chronology is applied to it.

Perhaps this is too rigid an interpretation of what Pétau has in mind. Where he refers to history as containing an 'orderly' account of individual years, this suggests a state of organisation going beyond the idea of 'pure potency' suggested by the language of matter and form.[8] Yet even if some latitude is applied to Pétau's characterisation of history in these passages, one thing seems clear: he is insistent on the logical and scientific superiority of chronology to history. Chronology as 'form' turns history into its true self by actualising its potential. Chronology, we might say, is history of a higher order.

---

[6] Petavius, *Rationarium* I 1, p. 1–2: *Historia rerum praeteritarum est narratio. Chronologia vero rerum earundem certis ex indiciis ad sua tempora temporumque partes accomodatio. Quare huic illa materiem, copiamque suppeditat. Haec illi vicissim formam, ac velut regulam adhibet.*

[7] Petavius, *Rationarium, Ad lectorem*, sig. a7r: *Habet hoc historia proprium, uti plenius rei gestae modum, ordinemque perscribat; nullis fere neque probationibus et argumentis, neque testibus, unde singulorum annorum ratio constet.* Later, he confirms that history *nec annorum singulorum exactam rationem instituit* (I 1, p. 2). For the *precise account* of the years, history is insufficient.

[8] Overall, it seems that Pétau moves from the strictly Aristotelian (non-dualist) understanding of matter as potency and form as its entelechy towards a duality of a material and its structure.

5.1 History as distinct from chronology — 59

To appreciate the implicit polemic in this presentation, we need to recall that chronology for the longest time was fundamentally an ancillary discipline to history intended primarily for beginners. Jean Bodin's influential *Methodus ad facilem historiam cognitionem* (1565), for example, proposed that chronologies ought to be studied *before* any of the more properly historical works could be read:

> First, then, let us place before ourselves a general chart for all periods, pure and simple, in which are contained the origins of the world, the floods, the earliest beginnings of the states and of the religions which have been more famous, and their ends, if indeed they have come to an end. [...] Conforming closely to this type are the works commonly called chronicles, characterized by spaces between the lines, brief indeed, but easy for beginners.[9]

The famous chronologies of the time, such as the *Chronicon Carionis* and Achilles Gassarus' *Historiarum et chronicorum totius mundi epitome* (Antwerp 1536) were 'obviously aimed at students in large part'.[10] As Anthony Grafton has pointed out, it was one of the main innovations of Scaliger's *De emendatione* to treat chronology instead as a separate discipline, a science in its own right, intended for scholars not beginners.[11]

Pétau clearly aligns himself with Scaliger in refuting the subordination of chronology to history, but it seems he seeks to go beyond a mere emancipation of the discipline from its function as history's handmaiden. Rather, he intends to reverse their order of priority. In the relatively few words he devotes to this problem, we can glean a remarkable demotion of history from a major part of the humanist curriculum to a proto-scientific endeavour riddled with so much uncertainty that only chronology can transform it into a proper science.

In making such an assessment, we have to tread carefully, acknowledging thoroughly traditional elements in Pétau's brief sketch of his understanding of history and its relation to chronology while also recognising the subtle ways in which this more traditional view is tweaked into a theory that seems at odds with a more conventional perspective in humanist historiography.

---

[9] Jean Bodin, *Methodus*, ch 2, p. 14: *Primum igitur communem velut omnium temporum tabulam, nudam illam ac simplicem nobis ad intuendum proponamus, in qua mundi origines, eluviones, rerumque publicarum ac religionum, quae magis claruerunt, initia duntaxat et exitus, si modo exitum habuerunt [...]. Ejusmodi fere sunt chronica qua vulgo feruntur, linearum spatiis distincta; brevia illa quidem, seu auspicantibus facilia.* English translation: Jean Bodin, *Method*, 21, amended.
[10] Grafton, 'Scaliger and Historical Chronology', 161, n. 24. Cf. also Chytraeus, *De lectione*, sig. B2r: *Sic in historiarum etiam lectione, principio epitome brevis seu chronicon ... cognoscendum est.*
[11] Grafton, 'Scaliger and Historical Chronology', 161.

To begin with, Pétau's description of history as the 'narration of past things' (*rerum praeteritarum narratio*) corresponds to one of the most well-established definitions of the discipline. It is found, for example, in Isidore of Seville's *Etymologies* and repeated over and over by the historians of the early modern period.[12] Moreover, the characterisation of history as 'blind' without chronology conveys an equally conventional sentiment. Bodin, for example, wrote that 'those who think they can understand histories without chronology (*sine ratione temporum*) are as much in error as those who wish to escape the windings of a labyrinth without a guide'.[13] Pétau's reference to what he calls a truism and a proverb, namely, that 'chronology is the eye of history' is also well attested among early modern writers, albeit usually in the form that chronology and geography are the 'two eyes of history'.[14] David Chytraeus, for example, cites this as a maxim in his introductory *De lectione historiarum recte instituenda* published in 1563:

> Therefore, students ought to know that knowledge of place and time, or topography and chronology, are, as it were, the two eyes of history.[15]

Casting our net more widely, we may furthermore observe that the reliability and truthfulness of historical sources was a *cause celebre* of early modern historical debate, often cast as the problem of a *narratio vera*.[16] In this, some sceptical influences reared their head which would make themselves felt more strongly half a century after Pétau's activity.[17] One such argument is from the *infinity* of historical material which may find an echo in Pétau's reference to the *copia* of history and in its characterisation as *materia*.[18] Overall, however, the problem of history's veracity is discussed in terms of the reliability of historical authors (were they contemporaries of the events they report or not?) and their embellishment of reports with rhetorical flourishes, such as lengthy speeches.[19]

---

12 Isidore of Seville, *Etymologies* I 41.1: *Historia est narratio rei gestae, per quam ea, quae in praeterito facta sunt, dinoscuntur.*
13 Bodin, *Methodus*, 323: *Qui sine ratione temporum historias intelligere se posse putant, perinde falluntur, ut si labyrinthi errores evadere sine duce velint.* English translation: Bodin, *Method*, 303. See also Chytraeus, *De lectione*, sig. B2r. Erasmus already called chronology history's *unica lux*: Grafton and Leu, 'Chronologia', 521 with n. 17.
14 Boyd Davis, 'May not Duration', 121–122.
15 Chytraeus, *De lectione*, B2v: *Deinde sciant studiosi: duos velut oculos historiae esse, locorum et temporum cognitionem, seu topographiam et chronologiam.*
16 Seifert, *Cognitio historica*, ch. 2.
17 Völkel, 'Pyrrhonismus historicus', esp. ch. 3.
18 Völkel, 'Pyrrhonismus historicus', 68–84.
19 On the contemporaneity of the ideal historian see Vossius, *Ars historica* I, p. 4–5; on speeches see Vossius, *Ars historica* I, XX–XXI, p. 96–111.

It is hard to imagine that these problems did not concern Petavius, but he is intent to sidestep them in the interest of what we might call a categorically different problematic. History as *merely* narration to him is aligned with an uncertain epistemic status. In *historia* we are by definition dealing with stories and reports without a scientific foundation. They may be more or less reliable, but the point Pétau is keen to emphasise is that *without chronology* they cannot at all be brought into a coherent and meaningful whole.

We can thus see that the metaphor of history's blindness and chronology as its eye does not, strictly speaking, work for his purposes any longer which is, perhaps, why he calls it 'very true but trite'.[20] While Chytraeus and many others used the formula to indicate the indispensability of chronology (and geography) for history, the point of the metaphor nevertheless invited the notion of chronology as an organ and thus as ultimately ancillary to history. For Pétau, by contrast, chronology is not an organ of history, but the principle of its full actualisation.

Consequently, history as *narratio* is not in the first instance plagued by more or less radical doubts regarding its veracity but by the limitations stemming from its approach to the sources. Thus far, Pétau's arguments for the priority of chronology over history mirror his earlier claims of chronology and computistics as the 'lower' parts of the science of time. As much as chronology needs to rely on astronomy and mathematics to attain to the dignity of a science, as much is history in need of chronology to accomplish the same end.

Perhaps the clearest sign of Pétau's intention to push the notion of history as narration to what would seem a counterintuitive extreme is his assertion that history offers neither proofs nor arguments. It is thus reduced to be the purely factual account of past events, but even as such, it will always lack precision without the support of chronology.[21] While this theory corresponds with the author's claim that history offers the matter to chronology as the form, it seems highly implausible as a description of what historians in Pétau's own (or at any time) have in fact done.

5.2 History as the realm of factuality

To understand Pétau's concept of history as pure narration without proofs or arguments, we need to look beyond the context of early modern historiography and

---

[20] See n. 5 (this chapter) above.
[21] See n. 7 (this chapter) above.

include philosophical debates of the time.[22] This seems warranted by Pétau's declared intention of basing his science of time on philosophical and specifically Aristotelian principles. While this nod to philosophy did not make him a scholastic, we have nevertheless seen how his use of the Aristotelian tradition in particular shaped his approach to the phenomena he investigated. It stands to reason that this stated intention applies to his definition of history too. After all, the categories he chooses are once again matter and form thus suggesting his continued operation within the non-dogmatic Aristotelianism which characterised his approach to the study of time in general.

At the end of Book II of his *De natura logicae*, the celebrated Aristotelian, Giacomo Zabarella (1533–1589) denied *historia* a place in the discipline of logic on the grounds that it is not an *ars* because 'it consists in the simple and pure narration of actions'.[23] An *ars*, however, requires an element that 'depends on our volition and is made by some human artifice'.[24] History is without those, the philosopher reiterates, but 'pure narration of actions lacking in any artifice except perhaps elocution.[25]

Here, it seems, we have precisely the nuance which Pétau too took care to emphasise, namely, that history was *only* narration without any theoretical addition (*omni artificio caret*, in Zabarella's phrase). This certainly flew in the face of attempts by several generations of humanist historians to establish their discipline as an *ars historica*.[26] Yet for Zabarella, an *ars* had to be productive of something: poetics, for example, produced literature and rhetorics, artful speech. This set the *artes* apart from *scientia* which, according to the Aristotelian philosopher was purely speculative.[27] In this scheme, the examination of causes or the explanation

---

22 We may recall here Pétau's strong emphasis on the philosophical character of his approach in the *Prolegomena* to vol. 1 of *De doctrina*. See Chapter 4 above (introductory section).
23 Zabarella, *De natura logicae* II 24, p. 67: *ars tamen historica non modo ab Aristotele, sed a nemine hactenus scripta comperitur; nec fortasse digna est in qua scribenda tempus conteratur; ea namque in simplici ac nuda rerum gestarum narratione consistit, quo fit ut nihil nobis fingendum, aut excogitandum relinquatur, ad quod facilius inveniendum praecepta de historia scribenda tradere, ac in artem redigere operae pretium fuerit.*
24 Zabarella, *De natura logicae* II 24, p. 67: *ars enim non dicitur nisi eorum, quae a nostra voluntate pendent et aliquo humano artificio fabricantur, cuiusmodi est logica tota et quaelibet eius pars, docet enim artificiose aliqua invenire, quibus tanquam instrumentis ad contemplandum, vel ad agendum iuvemur.*
25 Zabarella, *De natura logicae* II 24, p. 67: *At historia nil huiusmodi tractat, sed est nuda gestorum narratio, quae omni artificio caret, praeterquam fortasse elocutionis, quod quidem et alia eiusmodi quisque sanae mentis extranea et accidentaria ipsi historiae esse iudicaret.*
26 See the summary assessment in Seifert, *Cognitio historica*, 19–21.
27 Zabarella, *De natura logicae* I 2, p. 1–4.

of phenomena fell to the sciences, not the arts. The sciences, however, dealt with necessary and eternal things that were independent of human agency. From this perspective, *historia* seemed an odd fit for either of the two types of disciplines. It could not be *scientia* as it deals with human affairs, but it was not an art either as anything that was added to the report of *res gestae* would make it worse and not better.

In emphasising that *historia* merely provides a factual account, Zabarella drew on a development within Latin Aristotelianism that was started a century earlier by the Greek refugee scholar, Theodore Gaza who, after the capture of his native Thessaloniki by the Ottomans in 1430, fled to Italy where he became a key player in the humanism of the Quattrocento.[28] During a stay in Rome after 1450, Theodore produced new translations of some of Aristotle's writings, among them the *Historia Animalium*, as well as Theophrast's *Historia plantarum*. In his preface to this edition, Theodore explains the order of his collection as follows:

> The *Historia* occupies the first place and, as its name indicates, contains an exposition of the matter as it is (*quod est*) or as it may be, which the philosophical schools of our time conventionally call the 'because it is' (*quia est*). Subsequently, the books on the *Parts* and the *Generation* [sc. *of Animals*] present the cause for why things are as they are (*cur ita sit*).[29]

For Theodore, Aristotle's use of *historia* in this book title indicates an initial phase of fact-finding which, the translator explained, was needed before scholarship could proceed to the exploration of causes which is undertaken in subsequent works.[30]

By the early seventeenth century, this interpretation of *historia* was sufficiently established to make its way into collections of definitions but, characteristically, those produced by philosophers, not historians. In Rudolph Goclenius' *Lex-*

---

28 On Gaza see: Bianca, 'Gaza, Teodoro'. On what follows, see Seifert, *Cognitio historica*, ch. 3.
29 Theodore Gaza (trans.), *Aristoteles. De natura animalium*, [no pagination]: *Historia primum obtinet locum, atque ut nomen ipsum significat, expositionem continet rei, quod est, sive ut sit, quod scholae philosophorum nostrae aetatis, quia est dicere solent. Mox libri de partibus ac de generatione causa cur ita sunt declarant.*
30 This was indeed the traditional Greek understanding of the word ἱστορία as earlier Greek commentators on Aristotle, such as Simplicius had already pointed out: Simplicius [?], *In De anima*, ed. Hayduck, 7, 27–28: ἡ δὲ ἱστορία ἀντὶ τῆς ἐπιστήμης εἴληπται νῦν· ἴσμεν γάρ φαμεν ὅσα ἀκριβῶς γινώσκομεν (On Aristotle, *De anima* I, 402a3). Latin readers could find this etymology in Isidore of Sevilla, *Etymologies* I 41.1: *Dicta autem Graece historia ἀπο τοῦ ἱστορεῖν, id est a videre vel cognoscere.*

icon Philosophorum (1613) we find the following listed as the first definition under the entry *historia:*

> History signifies the knowledge of individuals, the exposition or description of a thing's 'what' (τοῦ ὅτι [...], *quod est*), as Gaza says. Hence [the title] *History of Animals*, that is, their simple description without demonstration.[31]

Note that neither Theodore nor Goclenius in the present passage pass judgment on history as a discipline. It took the work of a genuine historical sceptic to inscribe the conceptual logic of *historia*, which Gaza had established in the context of Aristotelian natural philosophy, into the interpretation of history proper. This happened in Francesco Patrizi's *Della historia: Dieci dialoghi* (1560).[32] Patrizi's teaching in the *Dialoghi*, it must be admitted, has been variously interpreted not least because it is difficult to ascertain the author's own viewpoint among the voices represented by the participants in his conversations.[33]

That said, whatever Patrizi's own position, it is clear that in his dialogues, a viewpoint finds expression that neatly illustrates the conceptual transition we are interested in. In Dialogue 5, Patrizi had allowed a certain Contile to voice radical doubts about the historian's ability to attain to truth with the conclusion that 'it is totally impossible for human actions to be known as they were actually done' partly because of the infinity of the material with which the historian has to operate.[34]

In the seventh dialogue, a novel solution to this problem is presented by Guidone who asserts that

> [...] all noble men of letters [agree] that the task of the historian is the narration exclusively of effects, whereas the search for the causes of any given thing is now the job of the philosopher.[35]

---

[31] Glocenius, *Lexicon philosophicum*, 626: *Historia significat singulorum notitiam vel expositionem seu descriptionem* τοῦ ὅτι *rei, quod est, ut ait Gaza; ut Historia animalium, id est, descriptio eorum simplex sine demonstratione.*
[32] Patrizi, *Della historia*.
[33] Franklin, *Jean Bodin*, 96–101; Seifert, *Cognitio historica*, 63–66; Völkel, 'Pyrrhonismus historicus', 75–80; Couzinet, 'History and Philosophy'. Couzinet gives an overview of the various interpretations, 'History and Philosophy', 64–72.
[34] Patrizi, *Della historia* V, f. 26r.: 'Non si puo adunque [...], in verun modo, [...] saper il vero delle attioni humane.' English text from Franklin, *Jean Bodin*, 100–101.
[35] Patrizi, *Della historia* VII, f. 41v: '[...] tutti i nobili letterati, che mestiere di historico sia, il raccontare gli effetti soli soli. Et che il ricercar la cagione di qual si voglia cosa, sia hoggimai ufficio da filosofo'.

5.2 History as the realm of factuality — 65

The difference between an account of facts or, as Guidone says 'of effects' and their causal explanation is here ascribed to the disciplinary boundary between history and philosophy. Elsewhere in the work, Patrizi consequently adopts the proposition that the historian is someone who 'takes note of effects without searching for, or speaking of, any causes'.[36]

Patrizi and Zabarella were near contemporaries, and even though their intellectual allegiances were diametrically opposed to each other in some regards, they both converge on the view that history is a merely factual description without the explanation of causes. Theodore's observation that the Greek term ἱστορία could sometimes be used for the 'simple description without demonstration' is thus broadened (or transformed) into the notion that history *as normally understood* offers merely a narration of facts.

Gerardius Vossius who in 1623 published what arguably was the most accomplished *Ars historica*, took Zabarella, the *philosophorum nostri seaculi princeps*, seriously enough to give pride of place in his great work to a refutation of his argument.[37] That said, this 'refutation' is ultimately rather qualified. Vossius concedes to Zabarella that *historia* was 'neither art nor science (*nec ars ... nec scientia*) and thus not a discipline at all.[38] Zabarella's mistake, according to the Dutch scholar, was merely his generalisation of what was true for *historia*, that is, the historical accounts of past events, to historical writing and analysis in general. It was the latter, which Vossius called in latinised Greek *historicé*, to which his own effort was dedicated.[39] From the outset of his work, therefore, he emphasised the need to distinguish *historicé* from *historia*: they relate, he argues, as poetics does to poetry, that is, it lays down rules 'for the production of history' (*ad conficiendam historiam*).[40] *Historicé*, Vossius argues, shows 'how annals or any

---

36 Patrizi, *Della historia* II, f. 8r: 'PATR[itio] [...] Ma mi dite, con qual nome chiamereste voi quello scrittore, il quale soli gli effetti vi contasse, senza alcuna cagione, a ricercarne, o dirne? BID[ernuccio]. Vorrete voi forse dire, che costui sia l'historico? PATR. Voi l'havete detto.'
37 Vossius, *Ars historica*, 6–8.
38 Vossius, *Ars historica*, 8: *Sic igitur statuimus, etsi historia proprie nec ars sit, nec scientia, atque adeo nec disciplina.*
39 Vossius, *Ars historica*, 8: *Tamen historicen esse artem, quippe quae circa universalia versetur, quod de historia dici non potest: ut quae occupetur circa singularia; idque eo fine, ut universalia praecepta inde colligantur, atque illustrentur. Laxe vero disciplinam ac scientiam esse, nemo negaverit, cum discatur et sciatur.*
40 Vossius, *Ars historica* I, p. 1–2: *Certe nihil magis extra controversiam poni debet, quam historicen differre ab historia; qua ratione distinguitur poëtica a poësi. Nam utraque disponit praecepta; illa ad conficiendam historiam, haec ad poësin.*

other histories should best be noted down in writing' (*quo, cum annales, tum quaevis aliae historiae scriptis optime consignentur*).⁴¹ This makes it an *ars*.

Despite this partial concession to Zabarella's dismissal of *historia*, Vossius clearly does not share the philosopher's reduction of history to pure narration. Rather, he explicitly includes the investigation of *causae ac consilia* among the tasks of history in contradistinction from annals which are restricted to a list of relevant events.⁴²

The different positions staked out by Zabarella, Patrizi, and Vossius against the backdrop of the Aristotelian terminology as elucidated by Gaza, help gauge Pétau's place in the debate. The cumulative weight of various authors' converging on the low theoretical status of *historia* proper explains the casual nature of Petavius' claim that *historia* was restricted to an account of individual years. What seems counterintuitive from today's perspective turns out to have been a remarkably attractive viewpoint at the time which could be shared by individuals with rather different theoretical commitments.

Not surprisingly, this understanding of *historia* gave rise to attempts to move beyond the uncertainty of historical knowledge. Zabarella alone seemed unconcerned about this problem and content simply to dismiss history from his philosophical edifice. Patrizi, by contrast, assigned to the Platonist philosopher the investigation of the 'invisible' causes of the visible events whose account was provided by the historian.⁴³ As for Vossius, he stipulates an *ars historica* as a kind of metahistory that can provide the theoretical foundations that history properly speaking lacks.⁴⁴

Compared with all these attempts, Pétau charts his own path with remarkable independence. Like Patrizi and Vossius, he postulates the need for another discipline to provide history's theoretical foundations, but he does not ascribe this task to either philosophy or to an *ars historica*. Instead, he proposes inscribing history into his own *doctrina temporum* for which, as we have seen, he claims the status

---

41 Vossius, *Ars historica* I, p. 5: *[Historice] generatim monstrat modum, quo, cum annales, tum quavis aliae historiae scriptis optime consignentur.*

42 Vossius, *Ars historica* I, p. 5: *Verumtamen quia, qui facere norit historiam, in qua exponantur causae, ac consilia, aliaque, quae haut requiruntur in scribendis annalibus, is quoque annalium conscribendorum artificium perspectum habebit: [...] eo omnis fere opera historices est de conficienda historia istiusmodi, qualem solam Semphronius Asellio historiae nomine dignabatur.* Cf. Semphronis Asellio, *Rerum gestarum libri* 1, fr. 1, ed. Peter, 179, 5–12.

43 Patrizi, *Della historia* VII, f. 41v: *Inquesta [...] che la cagione, in sua vera natura, anchor che cagione d'altro fatto sia, ella è però in se stessa, fatto. Et come tale, ella cade in narramento del'historico. Ma ella è dal filosofo, si come occulta nascosta cosa, & come cagione d'altra, investigata, & ricercata.* Cf. Seifert, *Cognitio historica*, 71–72.

44 See nn. 39 and 40 (this chapter) above.

of *scientia*. Not only, then, does Pétau offer chronology as a systematic umbrella to integrate history, he betters his competitors by promising that the formal shape provided by this doctrine will make history not merely *ars*, but *scientia*.

As for the latter point, we have previously seen that Pétau seems rather unconcerned with the technical debate around the division between arts and sciences.[45] Despite his clear insistence that the use of astronomy and mathematics gives his *doctrina temporum* the status of a science, he nowhere contrasts this with the possible interpretation of the discipline as an *ars* and, in fact, frequently refers to *chronologia* as an art.[46]

Ultimately, Pétau's proposal seems most directly opposed to the *ars historica* tradition. In keeping with his declared tendency throughout his work, he stays closest to early modern Aristotelianism where, as we have seen, the notion of *historia* as the realm of factuality was first introduced. At the same time, Pétau is not beholden to an orthodox or scholastic version of Aristotelianism but feels free to draw on this tradition in his pursuit of a rather idiosyncratic and original project.

5.3 Three methodological principles

Having clarified Pétau's distinction between history and chronology, we can now move to a consideration of how the two work together in his science of time. Pertinent in this connection are three methodological principles which, according to Pétau, jointly constitute chronology as a science: authority, demonstration, and hypothesis. These principles have rightly been singled out in previous scholarship as speaking to the remarkable character of Pétau's project.[47] Nevertheless, their interpretation is not entirely straightforward.

Pétau initially describes the three as follows:

> There are three kinds of principles that come together in Chronology. Some of them we understand by demonstration and through causes, such as those produced by means of the mathematical science. Others contain no knowledge of causes, and these are again divided into two. Some are accepted by hypothesis (ex ὑποθέσεως), for example, those the mathe-

---

45 See Section 4.4 above.
46 E. g. Petavius, *De doctrina, Prolegomena*, vol. 2, a2r: *Ars est altera reliqua, quam χρονολογικὴν vocant*.
47 Di Rosa, 'Denis Petau e la cronologia', 23–28; Wilcox, *Measure of Times Past*, 206–207; Hartog, *Chronos*, 131.

maticians call axioms. [...] Another kind, which is also the third kind of principles, relies on authority and testimony.[48]

In unpacking the ideas behind these principles, Pétau begins with the third one, authority. This, he writes, refers on the one hand to empirical truths, for example, that Jerusalem exists in our own time; that the Turks govern Constantinople or that the seas surround the dry land.[49] Pétau observes that everybody who has not witnessed these with their own eyes (undoubtedly the vast majority of Pétau's readers at the time) only knows about them by hearsay. Still these claims are often accepted as incontrovertibly true as people persuade themselves that they had all but witnessed them.[50]

The second kind of statements that fall into this category are reports found in the books of the historians or in other (historical) writings. These include the existence of quinquennial Olympic Games in Ancient Greece and that the city of Rome was founded by Romulus after the first Olympiad. The great historians, Thucydides, Xenophon, and Herodotus, we assume, lived even later and recorded those events from memory in their own time.[51]

Pétau does not here imply that these sources and testimonies are, or could be, unreliable. On the contrary, he insists that these generally accepted facts have their rightful place among the principles of chronology. Nevertheless, the author goes out of his way to draw a distinction between the value of information based on authority and the persuasiveness of arguments based on demonstration. It is

---

48 Petavius, *De doctrina*, vol. 2, *Prolegomena*, sig. a3r: *Tria sunt, quae ad chronologiam concurrunt, principiorum genera. Nam alia demonstratione, ac per causas intelligimus: ut quae media ex mathematicorum scientia promuntur; quaedam nulla causarum notitia constant: quae sunt rursus divisa bifariam; alia ex ὑποθέσεως assumuntur: cuiusmodi sunt ea quae mathematici postulata nominant. [...] Alterum genus, quod et principiorum tertium est, auctoritate, ac testimonio nititur.*
49 Petavius, *De doctrina*, vol. 2, *Prolegomena*, sig. a3r: *Ut esse hodie Hierosolymam; imperare Constantinopoli Turcam; terris maria circumfundi et huius generis alia nemo, qui non oculis viderit, aliter, quam fama et auditione cognoscit; et tamen sic ea persuadet sibi tamquam viderit: vix ut prioribus illis certius assentiatur.*
50 It is interesting that Pétau's authority-principle evidently includes *empirical* facts that are not technically 'historical'. This would support the interpretation of a 'drift' towards the alignment of *historia* with empirical reality in Pétau's work even though he does not seem explicitly to endorse this wider meaning of the term.
51 Petavius, *De doctrina*, vol. 2, *Prolegomena*, sig. a3r: *Ita quidam ab historicis, ceterisque scriptoribus accepimus adeo scripta constanter, ut de iis ambigere non dicam litterati hominis, sed vix hominis esse videatur. Talia sunt ista quoque: Fuisse quondam apud Graecos Olympiaca certamina, quae quinto quoque anno redibant, urbem Romani post institutas Olympiadas a Romulo conditam fuisse: Thucydidem, Xenophontem, Herodotum et id genus auctores illarum initio posteriores esse, tum eosdem res aetate sua, ac memoria gestas historiae mandasse.*

only through demonstration that we obtain necessary and incontrovertible knowledge. Whatever the value of authority and testimonies, 'sensible and prudent' people will ultimately be swayed by the full certainty of the causal argumentation included in the demonstration-principle whose conclusions, as Pétau pointedly adds, only the complete fool or the outrageously stubborn (*vel stolide vecors vel impudenter pervicax*) person will resist.[52]

In practice, this means that information gained from historical sources needs to be compared to the insights provided by astronomy and mathematics. This demand is in continuity with the line of argument Pétau had advanced at the beginning of the first volume. As seen in the previous chapter, he there argued that mathematics and astronomy are sciences in the strict sense of providing causal explanations whereas computistics and chronology on their own only report factual observations. If there is to be a science of time, then, the latter two 'lower' disciplines have to take the two 'higher' ones into their service.[53]

The principle from the initial section of *De doctrina* is, however, applied to chronology with a twist. The additional introduction of history as the 'matter' of chronology complicates the overall picture and turns chronology from a 'lower' discipline drawing on mathematics and astronomy into the place where the factual information provided by history is transformed into scientific knowledge with the help of the exact sciences. In presenting his three principles, Pétau does not, admittedly, reiterate the distinction of history and chronology as matter and form which he had introduced only a few pages earlier. Still, it is hard to avoid the conclusion that the authority-principle corresponds to his understanding of *historia*, as the purely factual basis of chronology. As history, in Petavius' interpretation, is restricted to the report of that 'which is' (*quod est*), so the authority-principle provides us with knowledge without proofs or evidence. Chronology thus receives the 'material' provided by history by means of its authority-principle and transforms it into scientific form through the application of the demonstration-principle.

In his initial introduction of the three principles, Pétau divided them into two categories: those with a knowledge of causes and those without such knowledge. The former group included the demonstration-principle, whereas the latter, he argued, was subdivided into two, namely, hypothesis and authority. Only a little later in the text, however, he holds demonstration and hypothesis together as

---

[52] Petavius, *De doctrina*, vol. 2, Prolegomena, sig. a3v: *postremum apud sanos et prudentes perinde vim demonstrationis obtinet. Quare quid potest eo ratiocinandi modo fingi certius, quod vel necessitate cogat ad assensum; vel cui persuaderi non possit, is aut stolide vecors aut impudenter pervicax habeatur?*
[53] Petavius, *De doctrina*, vol. 1, Prologemena 3.

ἀποδεικτικά and leaves the authority-principle alone as one that produces merely probable knowledge.⁵⁴ Does he contradict himself within less than a page?

The answer obviously depends on whether knowledge *demonstratione et per causas* and apodictic or necessary knowledge amount to the same thing. When we look at the demonstration-principle and the authority-principle, this could seem to be the case. Knowledge gained by demonstration is necessary because it includes an understanding of causes. With knowledge based on authority, it is the reverse. As it turns out, it is the hypothesis-principle that is more difficult to place; as such, it illustrates the small but vital distinction between the two classifications.

Pétau explains this last principle by comparing it to mathematical axioms. The example he provides is that the Christian era (*vulgo* identified with the year of Christ's birth⁵⁵) began 1627 years prior to the writing of his book. This, he argues, is true by postulation which means that 'although it is not based on demonstration but on use and convention, it can nevertheless not be denied by anyone and (therefore) serves for the completion of a demonstration'.⁵⁶ Pétau is thus not inconsistent on this point. The hypothesis-principle does *not* rest on any knowledge of causes because it is simply an assumption. At the same time, however, it falls in the category of apodictic or necessary knowledge because, *as what it is*, it cannot be denied.

Before probing more deeply into the meaning and significance of the three principles, it may be useful to adduce Pétau's own example for their use and their joint application. His test case is the beginning of the Peloponnesian War.⁵⁷ Thucydides reports that there was a solar eclipse in the first year of the

---

54 Petavius, *De doctrina*, vol. 2, *Prolegomena*, sig. a3v: *Ex his tribus colligatis invicem et connexis chronologicae probationes et argumenta ducuntur. Sed eorumdem principiorum priora duo ἀπο δεικτικὰ sunt, concluduntque necessario.*

55 Pétau regularly insists on the conventional character of this date. Cf. e.g. *De doctrina* VII 8, vol. 1, 622: *Et quamvis de vero anno Natali Christi digladientur inter se chronologi et uno anno, vel biennio, vel pluribus etiam annis vulgarem aeram antevertant; in aera tamen ipsa, sive annis Christi computandis, nemo ab altero discrepant.* Rationarium I 4, p. 16: *Quamvis enim de vero Natali Domini, hoc est anno, quo reipsa natus est, magna sit inter eruditos chronologos dissentio, omnes tamen annos Christi eodem modo computant.* See also his *De anno natali Christi diatriba*, in *Animadversiones*, 92–98.

56 Petavius, *De doctrina*, vol. 2, *Prolegomena*, sig. a3r: *Alia ex ὑποθέσεως assumuntur, cuiusmodi sunt ea quae mathematici postulata nominant. Exempli causa Christianam aeram, sive vulgare Christi annorum initium abhinc anno millesimo sexcentesimo, ac vicesimo septimo coepisse, postulatum est; quod etsi non demonstrarione, sed usu et consensione nititur, tamen neque negari ab ullo potest et ad concludendam demonstrationem valet. Sit enim eo concesso quod consequens est necessarium.*

57 Petavius, *De doctrina*, vol. 2, *Prolegomena*, sig. a3r: *Haec ut exemplis illustriora fiant, age propositum chronologo sit, ostendere, quoto ante nunc, qui supra millesimum ac sexcentesimum*

war. The fourth year saw Olympic games. Another solar eclipse occurred in the eighth year of the war, and a lunar eclipse in its nineteenth year.⁵⁸ There is further testimony not least from Xenophon who, like Thucydides, was a contemporary of these events.⁵⁹ Taken together, they point to the year 431 B.C. as the beginning of the war. Given we (i. e. Pétau's first readers) live in the year 1627 A.D., the war started two thousand and fifty-eight years ago.⁶⁰

This argument, Pétau continues, makes use of all three principles. To begin with, we apply the authority-principle according to which we have reason to trust Thucydides and Xenophon who wrote contemporaneous to these events and have a good reputation as historians.⁶¹ Next comes the demonstration-principle. For this, we compare the information provided by our historians with knowledge gained from astronomy, specifically the tables enumerating solar and lunar eclipses over the centuries. The varying eclipses mentioned by the historians can, in this way, be independently confirmed. In fact, this is *apodictic* knowledge which cannot be denied. Even the greatest despiser of chronology, Pétau notes, cannot find anything to object in information that is confirmed by astronomical calculation.⁶²

---

septimus et vicesimus aerae Christianae putatur annus, Peloponnesiacum bellum Lacedaemonios inter et Athenenses coeperit. Ad hunc ille modum colliget, itaque concludet [...]
58 Thucydides, *Historiae* II 28, 1 and IV 52, 1 for the two solar eclipses; VII 50, 4 for the lunar eclipse. See Stephenson and Fatoohi, 'Eclipses'. It seems less clear that Thucydides has much to say about Olympic Games in the 'fourth year' of the war, i. e. 427 B.C. He *does* devote an extensive discussion to the 420 B.C. games from which the Spartans were excluded: *Historiae* V 49–50. Hornblower, 'Thucydides'.
59 Note however that Pétau later in his work devotes a chapter to the proof that Xenophon's chronology is in fact erroneous: *De doctrina* X 28, vol. 2, 201–203. NB: in the 1703 edition, this section is X 29 since a new chapter 11 was added to the book.
60 Petavius, *De doctrina*, vol. 2, Prolegomena, sig. a3v.: *Quoniam enim Thucydides, qui aetate illa floruit, primo belli istius anno solem post meridiem obscuratum esse refert; tum in annum belli quartum Olympicum incidisse certamen; item anno belli octavo rursum defecisse solem; lunam vero anno decimo nono quo tempore Nicias Atheniensium imperator in Sicilia bellum administrabat; ad haec ultimo belli eiusdem anno, Olympiade nonagesima quarta solem defecisse testis est, qui eadem vixit aetate, Xenophon. (quorum testimonia libro ix ac x operis huius adduximus.) Ex his autem omnibus necesse fit, anno ante Christi epocham quadringentesimo tricesimo primo Peloponnesiaci belli primordia cecidisse. A Christiana vero ἐποχῇ ad hoc tempus anni sunt mille sexcenti viginti septem. Igitur anno abhinc bis millesimo et quinquagesimo octavo bellum istud incoeptum est.* Petavius offers a full discussion of this question in *De doctrina* X 27 [28 in the 1703 edition], vol. 2, 199–201.
61 Petavius, *De doctrina*, vol. 2, Prolegomena, sig. a3v.: *Nam Thucydidem et Xenophontem haud ista temere falsova scripsisse, tertium ad genus refertur.*
62 Petavius, *De doctrina*, vol. 2, Prolegomena, sig. a3v.: *Quod autem haec illorum vera esse nequeat historia, nisi anno ante communem aeram quadringentesimo tricesimo primo bellum inchoatum*

The hypothesis-principle comes into play when we translate the chronological information into a single system which, for Pétau, is the years before and after the beginning of the Christian era. His use of this system is in itself important, and we shall have to return to it. But for now, the main point is that this is indeed, as Pétau wrote earlier, analogous to a postulate. There is no need to calculate years in this manner, but doing so facilitates the integration of chronological information into a single system. In the present case, it is pertinent that the Christian era began in year four of the one-hundred ninety-fourth Olympiad (and at the end of year 754 *ab urbe condita*) since this permits aligning the chronological information from the ancient Greek sources with the Christian timeline.[63]

It is the use and application of these three principles, Pétau insists, that makes chronology scientific and reliable. Like medicine, chronology has both a rational and an empirical aspect. It needs both but while the latter is subject to interpretation and yields diverse opinions, the former produces certain knowledge. This can be observed in Scaliger, Pétau writes in something of a backhanded compliment: he often goes astray where he gives in to his own ideas and speculations, but 'did not go much wrong in the common periodical epochs extensively used by the best chronologers'.[64]

The same cannot be said of other recent chronologers who did not use the methodical approach (Pétau calls them στοχαστικοὶ et ἀμεθόδευτοι) and whose dates are therefore frequently inaccurate. Pétau demonstrates this by critically surveying the dates assigned to the beginning of the Peloponnesian War in the Chronicles of Johann Funck (1518–1566), Matthieu Brouard (1520–1576), and Gilbert Gén-

---

sit, ex astronomiae fontibus ac demonstratione concluditur. Itaque primum ad genus principiorum, quod est reliquis certius et ἐπιστημονικὸν pertinet. Adde iam lunae solisque, si lubet, cyclorum numeros, ut ab caeteris annus ille secerni possit; medio ex principiorum genere argumentatio illa proficiscetur. Libet ex his chronologicae artis irrisoribus quaerere quid in illa ratiocinatione desit ad probandi persuadendique firmitatem. Fac pertinacem quam voles adversarium; dum is ratione, communique sensu velit uti, quid habere contra potest quod obiiciat? An aut Thucydidem et Xenophontem sub illud tempus non fuisse, ac non ista scripsisse causari poterit; aut obscurationes istas siderum aliis annis potuisse contingere? Si prius dixerit, risum non tenebimus. si posterius, facile coarguemus.

63 Petavius, *De doctrina*, vol. 2, Prolegomena, sig. a3v.: Innumeaeri sunt in hac arte probationes eiusdem modi: ut quibus Olympiadum, Urbis conditaei, Nabonassari, ac similium intervallorum titulos atque epochas posteriore tomo stabilimus. E quibus insignis est et commemoranda maxime, qua vulgarem Christianorum aeram ostendimus cum anno quarto Olympiadis centesimae nonagesimae quartae; Urbis vero conditae septingentesimo quinquagesimo quarto exeunte concurrere.

64 Petavius, *De doctrina*, vol. 2, Prolegomena, sig. a4r: Itaque Scaliger, qui in ceteris fere lapsus est, quaecumque ex ingenio suo protulit et ad pervagatam aut ab aliis occupatam doctrinam addidit; in communibus et optimorum chronologorum usu contritis intervallorum epochis non multum aberrat.

ébrard (1535–1597), as well as the *Chronicon Carionis*.⁶⁵ As they mostly use the *Anno mundi* scale, he needs to translate from this into his own preferred B.C. system. Once the dates are thus made compatible, it appears that all these chroniclers – and Pétau is keen to emphasise their high reputation⁶⁶ – arrive at faulty results, usually by relatively few years. Does this mean that his own, scientific chronology is error-proof? Pétau does not claim that – after all, the practitioner of this discipline can misapply the methods or make other errors of judgments. It is, however, methodical and avoids self-contradiction.⁶⁷

For Pétau, then, the proof of his chronological pudding was in the eating: the results his principles yielded were reliable and scientific which, in turn, justified his approach. Nevertheless, it is important at this point to return to his theoretical frame and ask what the three principles tell us about his understanding of chronology and history before turning, in the next section, to the question of how they fit into his science of time.

An initial point to note is the specific way Pétau is using the language of causality here. When Zabarella, Patrizi, and Vossius disagreed on whether history included knowledge from causes, they clearly meant this to refer to the causes of the *res gestae* which are the main subject of the historical narrative. The question was thus whether the historian should simply relate the events leading, say, to the fall of the Roman Republic or whether they should also investigate the reasons why this chain of events took place. For Pétau, by contrast, the question is limited to the timing of events. This is why the causes he is interested in can be provided by mathematics and astronomy. The information we can glean from our sources in a non-scientific manner can and must be investigated, proved or critiqued by means of astronomical insight.

This indicates that Pétau's overall view of history is rather restricted. Vossius, in his discussion of the nature of history distinguished between history proper and annals: the latter, he conceded, had no need of *causae ac consilia aliaque*

---

65 Petavius, *De doctrina*, vol. 2, Prolegomena, sig. a4v–a5v. Cf. Funck, *Chronologia*, 70; Melanchthon, *Chronicon Carionis*, 263; Beroaldus, *Chronicum*, Index ad A.M. 3501 [no pagination]: *In Graeca confusa sunt omnia ista tempora. Itaque nec constat de tempore belli Medici neque de tempore belli Peloponnesiaci.* Génébrard, *Chronographiae libri*, 159.
66 On Funck: *ille inter ἀμεθοδεύτοις non ignobili loco et aetate sua celeberissimum fuit.* On the *Chronicon Carionis*: *a Philippo Melanchthone locupletum valde superiore saeculo et approbatione floruit.* On Brouard (Beroaldus): *non obscurus inter chronologos istos nominis.* On Génébrard (Genebrardus): *vir pius et apprime doctus.*
67 Petavius, *De doctrina*, vol. 2, Prolegomena, sig. a5v: *Ex quo fructus artis illius haud mediocris apparet, vel, ut desint caetera, magni faciendus: quod qui magistram illam sequitur, tametsi nonnunquam a veritate ipsa per imprudentiam aberret, sibi ipsi tamen consentaneus sit, nec antecedentia dogmata consequentibus elidat, quod est in eo genere turpissimum.*

but history itself did. Annals and chronicles were thus, for Vossius, part of history without exhausting the discipline.[68] Pétau clearly differs and considers the methodological confirmation of the chronological details contained in historical sources as the scientific perfection of history *per se*.

In this restricted sense, it can be said that 'history' and chronology are finely balanced out in Pétau's theory. Any information we have about the temporal dimension of a past event must be correlated with the 'cosmic clock', so to speak, in order to be meaningful. The most elaborate chronological reference in a source is worthless (as a chronological reference, that is) unless we can somehow tie it to the regularities of celestial movements. We can think again of the famous case of Ezekiel 1, 1 which seems to give so much chronological detail but ultimately fails to convey the knowledge about *when exactly* the prophet's vision took place.

At the same time, the most perfect astronomical knowledge would not get us any further without chronological references in our sources. Where human beings fail to record the exact point in time when some event took place, no scientific study of the starry skies will bring us any clarity either. We can think here of cultures that merely left behind mythical reports about their affairs. No scientific effort will arrive at reliable dates for events reported in such narratives.

Pétau and his fellow chronologists were, in fact, later criticised for this limitation. Giambattista Vico, for one, charged that they needlessly restricted human history to the relatively recent period during which it kept chronological records.[69] Whatever one wants to think of this criticism (and one might well want to defend the chronologists against this charge[70]), it seems clear that Pétau himself is less concerned about a part of human history without any chronological records and rather about people's willingness to take these records at face value.

---

68 See above (this chapter) at n. 42.
69 Vico, *La scienza nuova* II 10.2, ed. Rossi, 523: 'Perché, adunque, non ne incominciarono la dottrina donde aveva incominciato la materia ch'essi trattavano – perché incominciano dall'anno astronomico, il quale, come sopra si è detto, non nacque tralle nazioni che dopo almeno un mille anni, e che non poteva accertargli d'altro che delle congiunzioni ed opposizioni che le costellazioni e i pianeti si avessero fatti nel cielo, ma nulla delle cose che con proseguito corso fussero succedute qui in terra [...] — perciò tanto poco han fruttato a pro de princìpi e della perpetuità della storia universale (de quali dopo essi tuttavia pur mancava) i due maravigliosi ingegni, con la loro stupenda erudizione: Giuseppe Giusto Scaligero nella sua *Emendazione* e Dionigi Petavio nella sua *Dottrina de tempi*.' See also Henders, 'Vico's View'; Momigliano, 'Vico's Scienza nuova'.
70 As I have argued above, Pétau thinks that some awareness of calendric time is an essential feature of any human society. This thesis, admittedly, was facilitated for him by the assumption, shared generally at the time, that the chronological records in Scripture reached back to the origins of the cosmos, but it can be defended outside the biblicist frame of reference if understood as an anthropological theory about human social temporality. See Section 4.1 above.

His emphasis, in other words, is entirely on the insufficiency of 'authority' alone as a chronological principle and the need to align it with the evidence from astronomical and mathematical calculations. Accordingly, Pétau's main targets of criticism are those 'writers of chronologies and annals [...] who despise this craft and require nothing for the orderly instruction about the past except authority and the testimonies of the ancients'. However diligently these scholars use their methods, he observes, their work will never aspire to the height of science or amount to demonstration.[71]

The juxtaposition of an approach through authority alone and a truly scientific one speaks to a critical attitude towards tradition especially when considering that Pétau makes no distinction between secular and sacred chronologies.[72] Quite evidently, many of the chronological issues he faces are raised by biblical accounts. For all of them, his solution lies in the affirmation of a science that brings the demonstrative power of astronomy and mathematics to the information that can be gleaned from our sources, whatever their character may be.

That said, the kind of conflict a modern reader might expect between traditional insights and scientific resources is not necessarily on the cards in Pétau's system. The reason for this is a rather remarkable aspect of his authority-principle. As we have seen, the kind of knowledge contained under this principle falls into two main categories. On the one hand, there are empirical assumptions about the world which, even if we have not witnessed them first-hand, we accept as true. On the other hand, he refers to historical data which, once again, is accepted universally and could not be corrupted by anyone except through absence of common sense or propriety.[73]

---

71 Petavius, *De doctrina*, vol. 2, *Prolegomena*, sig. a2v: *Chronologiae et annalium scriptores aliqui sane pereruditi, sed huius artificii contemptores, qui ad ordinem temporum methodum aliam nullam requiri censeant quam authoritatem, ac testimonia veterum cum nonnulla in iis utendis ac digerendis industria: quae iudicii, potius et ingenii quam artis esse, scientia vero ac demonstratione constare nullo pacto videtur.*

72 Petavius, *De doctrina*, vol. 2, *Prolegomena*, sig. a2v: *Huic nos arti* [sc. chronologiae], *propter et varietatem ipsius infinitamque rerum copiam ac subtilitatem et ad usum litteratorum fructumque, commoditatem maximam: tum ut ardentioribus in ipsam maioris partis hominum, quam in caeteras eadem ex disciplina, studiis obsequerer, partem operis totius alteram seponimus; in qua τὰ ἱστορύμενα temporum, id est quaestiones omnes, quae de sacra pariter et profana historia solent oriri, arte ac via disputantur* (emphasis mine).

73 Petavius, *De doctrina*, vol. 2, *Prolegomena*, sig. a2r: *Haec et huiusmodi multa* [sc. examples of generally accepted historical claims] *quae longum sit percensere, quia confessa sunt inter omnes nec ea quisquam nisi vel communis sensus vel pudoris expers infitiari potest, merito inter chronologiae principia numerantur.*

In other words, the sort of information included under the authority-principle is what we would normally accept as secure and reliable empirical knowledge. As such, it does not seem in need of critical examination but only of 'demonstrative' and scientific confirmation. Pétau's example of the Peloponnesian War bears this out. The problem he seeks to address is *not* how to examine whether Thucydides' account is trustworthy, but only how it can be tied to astronomical knowledge about a solar eclipse at the time.

As we found previously in connection with his definition of history, Pétau leaves entirely to one side the many difficulties pertaining to the veracity of our historical sources.[74] It is certainly inconceivable that he was unaware of this problematic which loomed large in Scaliger's work which Pétau carefully studied. Moreover, the reliability or otherwise of ancient texts, not least ancient Christian ones, was a cause celebre in the historical and theological debates of his age.[75] Whatever his attitude to those issues, however, *in the present place* he decides to bracket them completely.

In fact, it is easy to see why he does so. The way in which, according to Pétau, astronomical insight confirms and solidifies the chronological information we gain from our sources only works if those sources are principally reliable. If, to use his own example, Thucydides was wrong about the solar eclipse in the first year of the Peloponnesian War; if it took place later during the campaign or if, for some reason, the historian had made it up for dramatic purposes, it would not help *at all* to summon astronomical tables about the timing of such eclipses in the fifth century. If anything, such a procedure would give a false air of scientific accuracy to a dubious chronological reconstruction.[76]

We should perhaps concede that Pétau's attitude to chronological information in our sources cannot have been naive. Even though he does not dwell on it in the present place, we therefore ought to ascribe to him the view that the authority-principle is meant to provide the best possible chronological information we can glean from the historical record. It is *this* information that then needs the kind of astronomical and mathematical confirmation to be scientifically accurate. This furthermore means that 'authority' in the present context does not have the

---

[74] See Section 5.1. above. In the present context it is intriguing that he *lists* Xenophon as a reliable historical witness but references a later discussion in *his own work* which mercilessly exposes the historian's errors. See n. 59 (this chapter) above.

[75] Cf. in the context of chronology, the Annian forgeries on which see Section 3.1 above. Perhaps the most iconic controversy of the time about a Christian text concerns the *Corpus Dionysiacum*. See Zachhuber, 'Dionysius', 537–538.

[76] As happens when Christian literalists use scientific methods to investigate, say, the date of Noah's Flood. See Pleins, *When the Great Abyss Opened*, ch. 10.

connotations it often has in theological, legal, or political contexts. It is not *some* authority that is beyond criticism but signifies the kind of knowledge we can obtain from the best available human reports about past events.

As much as one might accuse Pétau of underplaying the complexities of the authority- principle, one might also want to ask whether he does not overemphasise the significance of demonstration by means of astronomical data. In his example, it is admittedly crucial that the solar eclipse mentioned by Thucydides permits counter-checking his narrative against the astronomical clock. Is this not, however, a rather unusual case? Chronological information in our sources is not always, and not even arguably in the majority of cases, connected with unusual celestial phenomena. Very often, we are left with just the numbering of years or months that our sources provide, ideally with some reference to a dynastic scheme.

Could it be argued then that Pétau's insistence on the crucial importance of the demonstration-principle is exaggerated at least in practice? The response to this criticism would have to be twofold. First, while special celestial phenomena, such as solar or lunar eclipses are obviously not recorded for all historical events, a given event may still be relatable to another one for which such an occurrence is reported. If, say, we are only told that some event took place in the seventh year of the reign of a particular king, we may still have information that an eclipse took place in the year that king died. As long as we know how long he reigned in total, this helps solidify the chronological information we have of our initial event.

More importantly, perhaps, Pétau could argue that any calendric information makes use of astronomical data even if it 'only' contains references to years, months, weeks, and days. All of those are based on celestial phenomena, the twenty-four-hour solar day; the phases of the moon; the lunar or solar year; and so forth. Consequently, the knowledge and application of astronomy is required in principle even where no reference to unusual celestial phenomena is involved. Indeed, this is what Scaliger and Pétau himself undertook: to understand the astronomical bases of the different calendars used across cultures in order to make the calendars themselves compatible with each other.

In sum, Pétau's methodological principles demonstrate how he thought of the interplay of history and chronology in practice. His interest in historical records is from the outset geared towards the chronological information they contain. This information, he holds, is offered in the sources 'without causes' and therefore only as 'probable knowledge'. For this reason, Pétau chastises as inferior historians who rely exclusively on the authority of their written or oral sources. However trustworthy their authors, the scientific chronologist can only accept historical information if confirmed by reference to astronomical data.

5.4 Chronology, time, and the B.C./A.D. system

This leaves the question how Pétau's principles of chronology fit into his overall science of time. Clearly, his narrow understanding of history as a chronicle of past events is partly explicable from his premise that chronology is one aspect of societies' engagement with time. Recalling this overarching theme of his work helps put into relief his peculiar approach to history.

Moreover, the relationship between the time of historical chronology and the motions of celestial bodies studied by the astronomer, mirrors the duality of material and formal time introduced at the beginning of *De doctrina*. There is, admittedly, a source of confusion for the reader in Pétau's use of the language of matter and form in the two contexts. As we have observed before, 'material' time arises from the cosmic motions and is actualised by social time as its form. In the study of historical chronology, however, history is the 'matter' to chronology as form. Chronology, thus, actualises the account of time latent in historical sources; and it also actualises the cosmic time represented by the heavenly revolutions.

In this connection, Pétau's hypothesis-principle needs renewed attention. In what follows, I will argue that it is this principle that corresponds most evidently to the social time Pétau introduced at the outset of his work. This determination is important, as we shall see, because it permits adjudicating on some influential interpretations of Pétau's overall conception of time.

Let us begin by observing how closely Pétau here follows in Scaliger's footsteps. As we have seen at an earlier point in this investigation, Scaliger introduced the Julian Period as a postulated timescale by means of which every year would have its own character.[77] In this way, he gained a universal point of reference that worked independently of the type of calendar or 'epoch' used in any given historical source. Fundamentally, Pétau's hypothesis-principle is a precise methodological analogue to Scaliger's Julian Period (which the Jesuit is explicit in commending).[78] As much as Scaliger referred to the Julian Period as an 'invented' time, Pétau compares his hypothesis to mathematical postulates.[79]

---

[77] See above Section 3.1.
[78] Petavius, *De doctrina* VII 8, vol. 1, 622: *totidem* [sc. other dating systems] *sint opiniones annorumque summae: naturale illud rerum omnium, temporumque principium teneri non potest. Huius itaque loco Periodus Iuliana succurrit* [...].
[79] Petavius, *De doctrina*, vol. 2, *Prolegomena*, sig. a3r: *Alia ex ὑποθέσεως assumuntur, cuiusmodi sunt ea quae mathematici postulata nominant.* It is therefore misleading if scholars have occasionally presented Pétau's use of the B.C./A.D. system as a major deviation from Scaliger's practice. See Wilcox, *The Measure of Times*, 208; Feeney, *Caesar's Calendar*, 8.

The difference between the two comes into view in Pétau's immediate connection of hypothesis-principle with the Christian era. After initially proposing the date of 'the Christian era or *vulgo* the beginning of the years of Christ' (*Christianam aeram, sive vulgare Christi annorum initium*) as an example for what kind of postulate he has in mind,[80] Pétau subsequently 'translates' the ancient chronological date of the beginning of the Peloponnesian War directly into the number of years before the Christian era:

> We show that the [beginning of the] common era of the Christians corresponds to the fourth year of the one hundred ninety-fourth Olympiad, and the seven hundred fifty-fourth year *ab urbe condita*.[81]

Elsewhere, Pétau, having proposed to 'explain the benefit of the Julian Period and its remarkable use for the *doctrina temporum*',[82] transitions directly from a summary of the advantages of the Julian Period (cited in n. 78 above) to the benefits of the B.C./A.D. system:

> For (*etenim*) the Roman and Western Church retains the fixed and immutable epoch of the years of Christ on which all agree. For there is no one who would not count the present year in which I am writing this as the Year of Christ MDCXXIV.[83]

Pétau thus innovates in two ways. First, in addition to Scaliger's Julian Period, which was merely meant as an abstract, technical yardstick used by the chronologer to synchronise different dating systems, Pétau apparently advocates for a universal time which is also its own chronological system and can, as such, be used by specialists and laypeople alike.[84] Second, while in theory allowing for other 'hy-

---

80 Petavius, *De doctrina*, vol. 2, *Prolegomena*, sig. a3r. See n. 56 (this chapter) above.
81 Petavius, *De doctrina*, vol. 2, *Prolegomena*, sig. a3v: [...] *vulgarem Christianorum aeram ostendimus cum anno quarto Olympiadis centesimae nonagesimae quartae; Urbis vero conditae septingentesimo quinquagesimo quarto exeunte concurrere.*
82 Petavius, *De doctrina*, VII 7, vol. 1, 621: *Declaratur Iulianae Periodi fructus et ad doctrinam temporum usus insignis.*
83 Petavius, *De doctrina*, VII 7, vol. 1, 622: *Etenim Romana et Occidentalis Ecclesia fixam et immutabilem Christi annorum epocham retinet, in quam omnes consentiunt. Nemo est enim qui non hunc ipsum, quo haec scribimus, annum numeret Christi MDCXXIV.*
84 Petavius, *Rationarium* I 4, p. 17–18: *Quod si quis Periodum istam* [sc. Iulianam] *adhibere cunctabitur, poterit ad annos certo designandos ante aeram Christianam, in hanc ipsam summas suas dirigere, hoc est annos numerare, quot aeram Christianam id, quo de agitur, antecedat.* [...] *Hac nos putandorum annorum ratione hoc in libro saepius utemur, ne quid sit quod tyronem nostrum minus isti consuefactum periodo ab Chronologia studiis absterreat.* Cf. Klempt, *Säkularisierung*, 86.

pothetical' timelines to be universally applied in this sense, he seems to have in fact decided to prioritise the use of the *ante Christum natum / anno domini* scale.

Both these decisions have drawn the attention of previous scholars working on chronology and historiography of the early modern period. In 1960, Adalbert Klempt was the first (or one of the first) to discuss Pétau's role in the introduction of the B.C./A.D. system.[85] At the time, the target of his argument were scholars such as Oscar Cullmann and Karl Löwith who claimed that the *dual* chronological reference to years before and after the birth of Christ was a unique expression of a Christian temporality built around the Incarnation as the middle and centre of time.[86] Klempt points out not only the late arrival of the *ante Christum natum* chronology (as opposed to the counting of years *anno domini* which goes back to Dionysius Exiguus[87]), but also its rather pragmatic purpose. To him, the main motivation for its eventual introduction was the need for an alternative to the more traditional *anno mundi* system which was affected by the notorious issue of the large chronological variances between the Hebrew Bible and the Septuagint.[88] Pétau, in Klempt's estimate, sought to improve on Scaliger's Julian Period with a system that was easier to handle. In support of this claim, Klempt specifically referred to a comment in Pétau's *Rationarium* according to which the B.C./A.D. system would suit beginners who find the Julian Period too difficult to handle.[89]

Klempt, then, was keen to reject the notion that counting the years in both directions from the putative year of Christ's birth was profoundly theological. Rather, he saw the introduction of this system as part of a secularising tendency

---

85 Klempt, *Säkularisierung*, 85–89. Cf. his reference to Rühl, 'Die Rechnung', 628–629: '[...] zuerst [hat] Petavius diesen Gedanken [sc. of counting years B.C.] gehabt'.
86 Oscar Cullmann, *Christus und die Zeit*, 34. Cullmann knows that the 'gleichzeitige *Rück- und Vorwärtszählung von Christi Geburt an*' was only practised since the eighteenth century but argues that 'dieses Zeitschema der Zeit- und Geschichtsauffassung des Urchristentums entspricht' (italics in the original). Karl Löwith, who explicitly claimed Cullmann's authority for his treatment of 'the biblical view of history', went a step further: 'For the Jew, the central event is still in the future, and the expectation of the Messiah divides for them all time into a present and a future aeon. For the Christian, the dividing line in the history of salvation is no longer a mere *futurum*, but a *perfectum praesens*, the accomplished advent of Jesus Christ. With regard to this central event the time is reckoned *forward as well as backward*. The years of the history B.C. continuously decrease while the years A.D. increase toward an end-time.' (*Meaning in History*, 182).
87 Mosshammer, *Easter Computus*.
88 Klempt, *Säkularisierung*, 85.
89 Klempt, *Säkularisierung*, 86. Petavius, *Rationarium* I 4, p. 17–18. Citation in n. 84 (this chapter) above.

in early modern historiography.⁹⁰ Some later authors agreed with his interpretation of Pétau as a 'moderniser' but not with Klempt's 'pragmatic' interpretation of his chronological innovation. For Donald Wilcox and François Hartog, the use of the B.C./A.D. system was the most significant aspect of Pétau's contribution to Western ideas of time.⁹¹ It is on the basis of this particular innovation that Wilcox can claim that 'Petavius' system implied a single, continuous, and uniform time'.⁹² Hartog (who seems to rely entirely on Klempt and Wilcox for his own presentation) cites a sentence from *De doctrina* in which Pétau calls the 'first year of Christ's birth' the pivot (*cardo*) of times and 'a certain centre of history and chronology' from which 'the different numbers of years, like lines, extend into infinite distances of past and future, then again return from those to the same centre and are united together in it.'⁹³

In other words, it is Pétau's association of the hypothesis-principle with the Julian period and ultimately with the B.C./A.D. system which sustains the claim that he advanced a concept of 'absolute time' comparable in some ways to Newton's famous theory. It is thus arguable that at this point we touch on the core of Pétau's theory of time. How can the hypothesis-principle be inscribed into that theory? As we have seen, Pétau classifies it as not demonstrative but nevertheless providing necessary knowledge. This is because it has the status of an axiom. As such a postulate, it must surely be the product of human societies or cultures. In fact, Pétau repeatedly stresses the consensual nature of the hypothesis-principle. His preference for the B.C./A.D. system is at least partly justified by the existing acceptance of this Christian era in Western Christendom.

If then there is social time, that is, time constituted at the level of human society, the kind of dating system exemplified by the Christian era must be an important aspect of it. Given that Pétau had declared social time foundational for his new science, it is clear that the hypothesis-principle is key for his theory. Thus far, those scholars who have seen in it a crucial element of Pétau's theory, are right.

---

90 Klempt, *Säkularisierung*, 83: 'Die Notwendigkeit einer rechenhaft-präzisen und allgemeingültigen Datierungsweise für die geschichtlichen Ereignisse aller Zeiten wurde erst dann erkannt, als das neuzeitliche, welthaft-geschichtliche Denken an den altbekannten chronologischen Widersprüchen der biblischen Textüberlieferung so starken Anstoß nahm, dass der gewohnte Hinweis auf das gottgewollte Geheimnis der biblischen Unklarheiten zur Beschwichtigung der Fragen nicht mehr ausreichte.' Cf. also the title of the book!
91 Wilcox, *The Measure of Times*, 207–208; Hartog, *Chronos*, 131–132.
92 Wilcox, *The Measure of Times*, 207.
93 Hartog, *Chronos*, 132. See Petavius, *De doctrina* VII 8, vol. 1, 622: *Quamobrem primus iste Christi natalis annus cardo est temporum, sive quoddam historiae chronologiaeque centrum, a quo velut linea quadam, sic annorum varia summa in infinita spatia tam praeterita, quam futura propagantur et ab iis rursus ad idem centrum redeunt, in eoque coniunguntur.*

Recognising the relationship of the hypothesis-principle to social time does, however, raise one difficulty. Pétau insists that this principle is analogous to a mathematical axiom and is thus true by postulation. Yet social time, as we have seen, is not simply postulated but always depends on cosmic time. It seems to me that, despite Pétau's use of the language of postulate in connection with the hypothesis-principle, the same must also be true here. While the dating systems considered under that principle are axiomatic in the sense that the choice of any point of reference is conventional (Pétau emphasises several times that this is the case for the birth of Christ according to the Christian era[94]), in other respects they are dependent on the same astronomical knowledge that is also crucial for confirming information based on the authority-principle. Projecting chronological data from historical sources onto a universal timeline only makes sense if this timeline is itself solid in its astronomical foundations.

A further insight deriving from the association of the hypothesis-principle with social time concerns the plurality of dating systems. Pétau's reference to the Christian era as an *example* of how the hypothesis-principle can work is easily dismissed. As we have seen, his own use of this dating system seems to suggest that he considers it *de facto* normative. Despite this evidence from Pétau's own practice, however, there are reasons to take more seriously his claim that the B.C./A.D. system is merely *one* such scheme.[95] The hypothesis-principle, as such, only signifies that chronological data gained from historical sources and confirmed with reference to astronomical observations must always also be inscribed into a dating system that is shared by those meant to use this information. This would mean that, insofar as societies rely on a shared temporal framework, chronological work will inevitably always involve the application of Pétau's hypothesis-principle.

Using the Julian Period or indeed the B.C./A.D. system is therefore not the only way the hypothesis-principle can be applied, but the other major 'epochs' Pétau investigates in the wake of Scaliger serve the same purpose. This does not mean, however, that they are all of the same value. Rather, Pétau clearly thinks that Scaliger's introduction of the Julian Period has lifted chronology to an unprecedented and indeed an unsurpassable level. In other words, his own preferred dating system is best not only in the sense that recent chronological advances have improved on what was possible in the past. Rather, the progress

---

94 E.g. Petavius, *De doctrina* VII 8, vol. 1, 622; *Rationarium* I 4, p. 16. Full citation in n. 55 (this chapter) above.
95 Cf. again the introduction of the hypothesis-principles as a postulate at Petavius, *De doctrina*, vol. 2, *Prolegomena*, sig. a3r: *Exempli causa Christianam aeram* [...] *postulatum est*. See n. 56 (this chapter) above for the full citation of the passage.

achieved in the works of Scaliger and Pétau himself has led the discipline to perfection. Why this superlative? Pétau puts it distinctly by calling it *prima illa, longeque maxima commoditas* arising from the Julian Period 'that in this Period all Chronologies can be included' (*ut in eam Periodum chronologiae omnes includi possint*).[96] The operative word here is 'all' (*omnes*): the Julian Period proves its worth by its ability to comprehend and include all others. By embracing this one timeline, the totality of dating systems used in past (and present) human societies can be comprehended and included.

The scientific chronologist using the Julian Period, then, does not perform an operation without analogy in more traditional chronology, but they do so in a way that is not merely gradually but principally superior. As a consequence, the time which is the object of Pétau's scientific version of the hypothesis-principle is also not just different from and superior to the times objectified in traditional dating systems but is of a different kind. It is, in Pétau's mind, the consummation of more limited, traditional systems. Wilcox' phrase 'absolute' is correct in that sense.

It is more doubtful, however, that Pétau advocates the kind of absolute time to be found in Newton or even the concept previously encountered in Góis' Coimbrian *Physics* commentary.[97] To the extent that such an understanding of time is to be found in Petavius, it must correspond to his *material* time to which no limitations apply and which, for this reason, is in important ways unknowable.[98] The B.C./A.D. system, by contrast, is from the outset related to the structured and limited time human societies produce to facilitate their own temporality. While it provides the most comprehensive and the most universal such system conceivable, it nevertheless shares in the limitations that are essential, in Pétau's conception, for the kind of time that has been made operable for human individuals and their communities.

Writing history, for Pétau, is ultimately aimed at the integration of the mass of information we have in our historical sources into a shared timeframe. Pétau's focus on the chronicling of historical events can seem a narrow and restricted version of historiography, but to him such a characterisation probably missed the point. The problem as he saw it was not the possible exclusion of valuable historical insight from the scope of the scholar's work, but human inability to cope meaningfully with the wealth of material contained in our sources. In its raw form, this material threatens to inundate and overwhelm, and the primary task for the historian is therefore to organise and structure it. History, if understood

---

96 Petavius, *De doctrina* VII 8, vol. 1, 622.
97 See Section 4.2. above.
98 See Section 4.3. above.

as a conception of societies' past, only emerges in this process. Chronology, and especially the hypothesis-principle, creates the temporal order into which the historical material can then be projected. The result is a consciousness of our past as a time filled with historical events that occurred in a specific order and can be remembered as such.

5.5 Conclusion

It was the task of this chapter to consider Pétau's understanding of history within his project of a science of time. The starting point of this investigation was the author's distinction between history and chronology. Pétau's definition of *historia* as mere matter, an account of events without causal explanation, seemed initially counterintuitive but could be shown to fit into a broader debate about the nature and the character of historiography in early modernity. Pétau sides with some philosophers against the humanist mainstream in his low esteem of the disciplinary dignity of history as such. History, to him, is in need of chronology to be turned from an accumulation of random facts into a proper science.

The next step was a closer analysis of the process by means of which this happens. From a study of the *distinction* Pétau draws between history and chronology, the investigation thus transitioned to their methodological interaction. In explaining the transformation of historical datapoints into solid chronological knowledge, Pétau follows the logic established in his initial exposition of his *doctrina temporum* making the application of mathematical and astronomical insights key for the advancement of history to properly scientific chronology.

Pétau's emphasis on the need to control historical insight by reference to astronomical data could give the impression that he is less concerned with historical criticism as commonly understood. As a matter of fact, his use of the term 'authority' in *De doctrina* might suggest a rather naïve view of the character of information that can be gleaned from historical sources. Such a view should not, however, be ascribed to Pétau whose affirmation of historical critique is clear from his more historical works even if it is bracketed in the context of his work on time.

The connection between Pétau's approach to history and chronology, and his overall theory of time became even clearer once his so-called hypothesis-principle was included in the analysis. This principle, I argued, represents Pétau's notion of social time in the context of chronology. In order for human societies to have an awareness of their past, historical records need to be integrated into a constructed theoretical framework; in practical terms this means fitting them into a dating system. Pétau aimed at the most comprehensive such system. Taking his cue from Scaliger's Julian Period for maximal scientific accuracy, he advocates for pragmatic purposes the use of the B.C./A.D. scale. In this way, Scaliger's abstract tool of the chronologer became a practical dating system commendable to everybody's use. Moreover, this universal dating system was based on the accepted calendar of Western Christendom.

In adjudicating the significance of this step, recent scholars have sought to resist the intuitively plausible assumption that Pétau's adoption of the B.C./A.D. system was the expression of a theological or religious agenda. Instead, they have emphasised his pragmatic concerns and insisted that his establishment of an 'absolute' system of dating was in tune with contemporaneous scientific and philosophical advances.

While it is undoubtedly legitimate to express reserve towards a simplistic identification of a religious motive in Pétau's innovation, there is a danger that a rather obvious theological dimension in Pétau's presentation is lost. As I will argue in what follows, attending to this dimension helps add important further clarification to Pétau's theory as well as his understanding of his own time and place, and their importance within human history.

To this end, I shall now turn to those sections in *De doctrina* from which it is clear that social timekeeping for Pétau had an ineluctably religious aspect before examining the even more overtly theological characterisation of time in his dedicatory letter to Cardinal Richelieu.

6 Social time as religious time

It has been argued in an earlier chapter that for Pétau societies are by nature temporal: they inscribe their communal existence into a calendric and chronological framework and, by doing so, constitute subjective time.[1] For any individual to orient themselves in time, the prior existence of social temporality is necessary. Any conception we may have of time or duration, and certainly any *quantifiable* notion of time is prefigured for us in the culture within which we grow up.

Pétau can occasionally speak of this social temporality in rather pragmatic terms. For example, he defines his science of time as inquiring into the conditions and properties of time 'insofar as it can be applied to the use of human beings' (*ut ad usum transferri potest hominum*).[2] Similarly, he describes it as being concerned with that aspect of time by means of which 'it can be made suitable to the human race's civil use and management.'[3]

This kind of expression must not, however, occlude the fact that for Pétau the social handling of time was closely linked to religion. For societies to have temporality is ultimately the same as saying that they have religion. This is first of all the case because, in Pétau's historical analysis, the calendric systems societies invent are most fundamentally needed to fulfil their religious obligations. This is clear from his brief historical account of the emergence of the science of time in Chapter 2 of the prolegomena to *De doctrina*.[4]

Pétau there sets out with some comments about time measurement among the Greeks. His source is the *Elementa astronomicae* by Geminos, a Hellenistic astronomer of the first century B.C.[5] Geminos' introduction was important for Pétau's work throughout as indicated by the fact that Pétau produced his own improved edition of the Greek text as part of his work on *De doctrina*.[6] According to Geminos, as Pétau elucidates, ...

---

[1] See above, Section 4.3.
[2] Petavius, *De doctrina*, Prolegomena 3, vol. 1, sig. i1r: *Quam sane χρονικὴν deinceps nominemus, ac definiamus ita si lubet, ut sit scientia quaei temporis, ut ad usum transferri potest hominum, conditiones, ac proprietates inquirit.*
[3] Petavius, *De doctrina*, Prolegomena 1, vol. 1, sig. e3v: *Reliqua est quarta, postremaque ratio [...] cum in tempore id spectatur unum, quatenus ad civilem humani generis usum, tractationemque conformari potest.*
[4] Petavius, *De doctrina*, Prolegomena 2, vol. 1, sig. e4r–e5v.
[5] Geminos, *Elementa*, ed. Manitius. English translation: Evans and Berggren (trans.), *Geminos's Introduction*. On Geminos and his work see the Introduction, Geminos, *Elementa*, 1–110.
[6] Petavius, *De doctrina* 1703, vol. 3, 1–39. Cf. the description of Pétau's edition in Manitius' *Praefatio* to the Teubner edition, p. iv–v.

> [...] the most ancient Greek states sacrificed as commanded by their oracles, κατὰ τρία (according to the threefold order) and κατὰ τὰ πάτρια (according to the ancestral rite). In this way, they maintained the right order of days, months, and years in their religious holidays in order to celebrate them at the same dates which they had received from their forebears.[7]

This explains why they used the solar cycle for their years, but the lunar cycle for days and months. For only the use of the solar year ensured that their festivals always fell into the same season of the year. For the months and days, things were different as here the phenomena of new moon and full moon predominated. The days finally were named after the phases of the moon.[8]

It was the need to synchronise these different calendars which set off, in Pétau's interpretation, the initial attempts at a science of time in Ancient Greece. Repeated attempts were made to find the best cycle permitting the use of the different systems. For this, an ever-improving astronomy was necessary clearly anticipating what, in Pétau's estimation, continues to be the case in his own time.

> So much for the Greeks, from whom astronomy, like all the cultivated and developed arts – and the science of managing time which is connected with it – soon spread to the other nations and states.[9]

All the details in this account are from Geminos, but whereas the Hellenistic author introduces the reference to the sacrificial requirements underlying the invention of the earliest Greek calendars rather as an anecdote, for Pétau this fact is much more important.[10] If this is not already evident from the way he places it

---

7 Petavius, *De doctrina*, Prolegomena 2, vol. 1, sig. e4r: *Antiquissimae Graeciae civitates, cum ab Oraculo praeceptum esset, uti κατὰ τρία (tribus observatis) et κατὰ τὰ πάτρια (patrio ritu) sacrificarent; hoc est ut mensium, dierum, annorumque ratio in sacris haberetur, atque ut iis ipsis, quae maioribus acceperant, temporibus illa celebrarentur.*
8 Petavius, *De doctrina*, Prolegomena 2, vol. 1, sig. e4r: *Sic istud interpretati sunt, quasi annos ad solis, menses ac dies ad lunae conversiones exigi oporteret. Annos porro ad solem dirigi, nihil aliud esse, quam sacrorum caeremonias in easdem perpetuo anni tempestates incidere, ut quae vere, aut per aestatem, vel hiemem instaurari fas esset, numquam alienis temporibus agerentur. Quod quidem fieri nulla ratione poterat, nisi solstitiorum, aequinoctiorumque cardinibus constantes et immobiles in anno sedes attributae fuissent. Nam ad lunam accommodare menses id esse, tam ipsorum initiis coniunctiones cum sole lunae, quam mediis iisdem plenos illius orbes congruere. Ex quo tertium redundat, ut ex lunae diversis illuminationibus, prout accedens ad solem, ab eoque digrediens, lucis incrementa vel defectus accipit, dierum quoque vocabula sumantur.*
9 Petavius, *De doctrina*, Prolegomena 2, sig. e4v.: *Haec de Graecis, a quibus, cum omnes elaboratae propagataeque artes, tum astrologia, et, quae cum ea coniuncta est, gubernandorum scientia temporum, paulatim reliquis gentibus ac civitatibus infudit.*
10 Geminos, *Elementa* VII 6–7, ed. Manitius, 102, 7–14: Πρόθεσις γὰρ ἦν τοῖς ἀρχαίοις τοὺς μὲν μῆνας ἄγειν κατὰ σελήνην, τοὺς δὲ ἐνιαυτοὺς καθ' ἥλιον. Τὸ γὰρ ὑπὸ τῶν νόμων καὶ τῶν χρησμῶν

at the outset of his own summary, its significance becomes clearer as the section progresses. For after some comments on the (inferior) traditional Roman calendar and its reform by Julius Caesar, Pétau turns to the Christian contribution to the science of time.[11]

Here, he dwells on the debates about the date of Easter. Once again, it is thus a religious ritual that necessitates calendric adjustments. He emphasises the urgency of the remedial action: the Council of Nicaea in 325 was 'summoned no less for the establishment of a rule of ecclesiastical time than for putting out the Arian conflagration', as Constantine testifies in his rescript, along with Eusebius and some others.[12] This occurred, Pétau argues, by means of a principled reflection on the dating of this event. The Alexandrian computation was declared universally binding because of their superior astronomical and calendric expertise, and the bishop of that city received the right to determine the date of the festival on an annual basis.[13]

Thus, while the need to observe religious rituals in a timely fashion led to calendric innovation, the consultation of the most advanced science of the day was indispensable for it to succeed. Where the latter is lacking, problems ensue as Pétau subsequently details where he explains the reoccurrence of the Easter controversy in the early Middle Ages which he blames on various factors including the adoption of a 'poorly understood nineteen-year cycle' and ignorance of Greek among the Latins at the time.[14]

παραγγελλόμενον, τὸ θύειν κατὰ [γ', ἤγουν] τὰ πάτρια, [μῆνας, ἡμέρας, ἐνιαυτούς], τοῦτο διέλαβον ἅπαντες οἱ Ἕλληνες τὸ τοὺς μὲν ἐνιαυτοὺς συμφώνως ἄγειν τῷ ἡλίῳ, τὰς δὲ ἡμέρας καὶ τοὺς μῆνας τῇ σελήνῃ. Note that the phrases in square brackets are included in Pétau's edition: *De doctrina* 1703, vol. 3, 18.

11 Petavius, *De doctrina*, Prolegomena 2, vol. 1, sig. e5r: *Eadem in nostris, hoc est Christianis, religionibus moderandis doctrinae temporum et progressio, et opportunitas perspici potest.*

12 Petavius, *De doctrina*, Prolegomena 2, vol. 1, sig. e5r: *Huius enim conventus habendi consilium non minus ad constituendam in Ecclesia temporis disciplinam, quam ad Arianum incendium restinguendum pertinuisse, Constantinus ipse in rescripto suo, et Eusebius, aliique testantur. In eo Patrum consessu ita decretum est, ut eodem omnes Ecclesiae tempore Pascha celebrarent.*

13 Petavius, *De doctrina*, Prolegomena 2, vol. 1, sig. e5r: *Nihil hoc erat, nisi illius ipsius temporis certa et communis quaedam extaret ratio. Haec vero uti quam fieri posset exquisitissima traderetur, datum Alexandrinis negotium est, qui siderum peraeque, ut ordinandorum temporum scientia, erant maiore quam exteri, ut illorum praesul legitimum celebrando Paschati diem quotannis incideret.* Cf. Leo the Great, ep. 121, 1, *Patrologia Latina* 54, 1056.

14 Petavius, *De doctrina*, Prolegomena 2, vol. 1, sig. e5r: *Ac Latinorum praecipue secta in Galliis, Hispanis ac Britannicis insulis quamplurimorum usu patrocinioque floruit, quos et Quartadecimanos appellarunt. Quorum ex praepostera cycli decemnovennalis institutione, necnon Graecarum inscitia litterarum, et Rufini* παρερμηνείᾳ *temere susceptus error, ac longo Catholicorum ratiocinio saepe fractus et convictus, aegre tandem a Beda, ceterisque moderandorum in Ecclesia temporum*

The culmination of this narrative inevitably arrives with the calendar reform under Pope Gregory XIII. Pétau does not go into the details of the religious problems underlying the need for this reform although it is clear that once again the problems of celebrating Easter at the correct time in the year was a major concern. Instead, he focusses on its accomplishments and its overall perfection in resolving current and future calendric difficulties as well as proving the benefits of a properly applied *scientia temporum*.[15]

The pre-history of his own science of time suggests, then, that Pétau considered social time fundamentally religious. As much as societies inscribe themselves into a calendric system based on the regular movements of the heavenly bodies, this process is nevertheless always closely connected with religion.

Pétau's position on this matter becomes even clearer from his dedicatory letter to Cardinal Richelieu. Time, he there points out, has traditionally been seen as a god and, more specifically, the father of all other gods.[16]

> From this, I explain the fact that the office of ruling and administering time once fell among all nations exclusively to priests and High Priests (*pontifices*) so that, by assent of common nature, to them is assigned the governance of both the cult of sacred things and of time, which is the most sacred of all things. This all the more as among all the rituals in which a heavenly deity is venerated, the observance [*religio*] of time, that is, of festal days and holidays is not the smallest one.[17]

Pétau here continues with a long and intriguing list of nations and their priests who allegedly had this kind of office: the Chaldeans among the Babylonians,

---

consultis opprimi obliterarique potuit. Cf. Ohashi, 'Theory and History'. On Rufinus' 'mistranslation' cf. Petavius, *Animadversiones*, 193–196 and *De doctrina* VI 11, vol. 1, 564D.

15 Petavius, *De doctrina*, Prolegomena 2, sig. e5v: *Quo in genere praeclara est et ad genus hoc disciplinae praedicandum ampla ac magnifica Gregoriani Kalendarii descriptio: qui labantis et inclinantis anni ruinas non solum fulcit, sed huic etiam integritatem incolumitatemque contulit: immo vero, quod hactenus deerat neque veterum quisquam consequi potuerat, perpetuitatem et constantiam. Huius artificii singularis cum elegantia et facilitate iuncta subtilitas, cum omnia hoc in genere veterum omnium inventa superavit, tum ad summum hanc, cuius foetus ac fructus est, evasisse regendi scientiam temporis ostendit.*

16 Petavius, *De doctrina*, Epistola sig. a3v-a4r: *Hinc veteres illi, sacris fabulis, non solum Deos inter adscriptum tempus; sed etiam factorem ac parentem ceterorum esse docuerunt.* Further on the Kronos-chronos etymology and its use in Renaissance and early modern thought see below Chapter 7 at n. 13.

17 Petavius, *De doctrina*, Epistola, sig. a4r: *Ex quo factum illud interpretor: uti regendi gubernandique temporis munus apud omnes populos soli quondam sacerdotes ac pontifices obierint; ut iisdem, communis naturae suffragio et sacrorum cultus, et rei omnium sanctissimae temporis procuratio concessa sit. Praesertim cum inter omnes ritus, quibus caeleste numen colitur, non minima sit temporis, hoc est feriarum ac festorum dierum, religio.*

the Brahmans among the Indians, the hierophants among the Egyptians, and so forth.[18] He claims that Julius Caesar acted in his capacity as Pontifex Maximus when he instituted his calendar reform – something he does pointedly not say in his more scientifically oriented prolegomena.[19]

While there are thus differences between the rhetorical flourish of the dedicatory letter and the sober and factual language of the prolegomena, their overall drift is clearly the same. Both accounts, notably, lead up to Pétau's celebration of the Gregorian reform which, in the letter, he inevitably credits to Gregory as the Pontifex Maximus.[20] Both make it clear that social timekeeping is not a 'secular' but a sacred affair. It is part and parcel of a nation's religion, the way this nation relates to God or the gods. Its administration therefore is inseparable from the office of priests, the people that is, whose overall task is the mediation between the earthly community and the deity.[21]

That said, priests only discharge this responsibility well if they are informed by the best science of the day. This is how Pétau understands the history of his discipline, as a series of religiously prompted and scientifically executed improvements of humanity's calendric and chronological knowledge. The apex of this development has arrived in his own time. Pétau, as we have seen, is not modest in this regard but believes that the accomplishments of the Gregorian reform and the kind of scientific chronology instituted by his nemesis, Scaliger, have led this history to its ultimate conclusion.[22] It is by the same reckoning that he believes it is

---

18 Petavius, *De doctrina, Epistola*, sig. a4r: *Igitur apud Babylonios, Chaldaei; apud Indos, Brachmanes; apud Aegyptios, Hierophantae; apud Romanos, Pontifices; apud Gallos nostrates, Druidae; apud Iudaeos, Sacerdotes popularibus fastis ordinandis, summo iure praefuerunt.*
19 Petavius, *De doctrina, Epistola*, sig. a4r: *Romae quidem prae ceteris tam infinita penes hunc ordinem potestas illa fuit, iidem ut arbitrio suo producerent aut contraherent annorum spatia, et eam labem pedetentim consciscerent, quam Iulius Caesar eadem Pontificii magistratus auctoritate compressit.*
20 Petavius, *De doctrina, Epistola*, sig. a4r: *Omnem quidem fastorum corruptelam, ac perniciem vetustate corroboratam et in dies grassantem, Gregorius decimus tertius Pontifex Maximus pari felicitate ac Christiani orbis approbatione sustulit.*
21 Petavius, *De doctrina, Epistola*, sig. a4r: *Age num illud est obscurum de temporis religione populorum omnium iudicium, quod annales nulli admodum praeterquam sacerdotes attigerunt? De iis loquor annalibus, qui publica fide perscribebantur.* The last sentence curiously recalls a reference to 'Megasthenes' in Bodin, *Methodus*, 42: *Neque tamen, inquit, omnes probandi sunt, qui de regibus scribunt, sed solum sacerdotes penes quos publica fides est annalium suorum* [...]. Bodin in his turn cites from Annius' *Antiquities* XI, *praefatio*.
22 Petavius, *De doctrina, Prolegomena* 2, sig. e5v; 3, sig. i1v.-i2r.

his mission to inaugurate a novel form of chronology, a science of time, which as he asserts has not hitherto existed.[23]

If, then, social timekeeping is an important aspect of the nation's religion; and if, furthermore, Pétau clearly believes a science of time can become reality in his own age, authored by himself, a Jesuit priest, and dedicated to a cardinal, the conclusion can hardly be avoided that the implicit cultural premiss of his approach is his deep conviction of the perfection of religion in his own, Tridentine Catholicism.

To conclude, as some scholars have done, that Pétau's understanding of time is secular or in other ways divested of his own, Christian commitment is therefore misguided. For Pétau, the accomplishment of a truly scientific chronology, a science of time, and the production of a universally applicable dating system are all evidence of the superiority of the religious status quo in early modern Catholic Europe. And yet it is easy to see how such a misunderstanding arose. Pétau's articulation of this superiority was not confessional in a modern (or perhaps also: a more traditional) sense. He did not appeal to the significance of any doctrinal tenets of the Christian faith as proof of its unique value. Klempt was certainly right to maintain against Löwith and others that the introduction of the B.C./A.D. system was *not* motivated by the belief in the universal reach of the Incarnation, as testified by Pétau's apparent openness regarding the date of the birth of Christ.[24]

The unique place of modern Catholicism in the history of religions lies, rather, in its perfect collaboration with modern science and its ability to use scientific insights to generate a calendar and a chronological system superseding all previous ones in such a way that further progress seems unnecessary or even impossible.

With regard to historical chronology, Pétau states emphatically that the same methodology applies to both secular and sacred texts. Both are transmitted to us as historical documents and thus belong to the mass of data that needs scientific investigation in order to be transformed into historical knowledge proper. It is well known that Pétau was an active participant in contemporaneous debates about the year of Christ's birth, and Oudin, in his biographical sketch, mentions that some of his fellow Jesuits were uncomfortable with his critical stance on this issue.[25]

---

23 Petavius, *De doctrina*, Epistula, sig. a5v-a6r: *Hanc nos incredibili multorum annorum studio ac labore pervestigatam, et confectam opere isto complexi sumus [...]. [...] cum rei sacrosanctae, temporis inquisitionem, tractationemque contineat, et hoc ipsum non vulgari communique modo, sed subtili et ad intelligendum perdifficili tradatur, cumque vel re ipsa, vel — ut dicam modestius — opinione nostra, nunc primum in absoluta scientia redacta formam, quasi nova quadam disciplina prodeat.*

24 Klempt, *Säkularisierung*, 81–82.

25 Oudin, 'Denis Petau', 112: 'Quelques-uns de ses confrères voulurent lui faire un scrupule, de ce que sur l'arrivée de la naissance de Jésus-Christ, il avoit abandonné le sentiment de Baronius,

Yet Pétau made no apologies for his approach to such matters. In his prefatory *Ad lectorem* to the *Rationarium*, he saves some of his most acid words to chastise a particular kind of chronologist:

> Among the many chronologies that exist to this day, none is worse than the one that assumes for itself, with the greatest unworthiness, the glorious name of sanctity (*sanctitatis* [...] *nomen*). I believe [it does so] to fend off human inquisitions and judgments by pretending to be a sacred thing, even a point of religion. And yet it is of the kind that no learned person would approve, and only the ignorant can praise.[26]

While it is arguable that Pétau here targets one specific work, the *Saincte chronologie* published by Jacques d'Auzoles Lapeyre in 1632, it remains significant that he expresses his reproach in such general terms.[27] The quality of scientific work, the reader is invited to understand, depends on the rigorous application of methods and the careful parsing of sources, not on an appeal to religious faithfulness. Ultimately, the solution to chronological or historical attacks against religion is not the insulation of matters of faith from scientific enquiry but the rigorous pursuit of such enquiry.

Religion, then, is not juxtaposed with science, but the two work hand in hand. In fact, throughout the history of calendars, religious authorities utilised scientific insight to the extent that they could. Looked at from the perspective of Pétau's science of time, the history of the relationship of science and religion is thus one of mutual dependency and collaboration. If there is progress in this history, it consists in the ever-improving symbiosis of the two.

While one has to admit that Pétau does not yet have the kind of evolutionary concept of religion that became *en vogue* a century after his death, his theory seems to point in this very direction with the upshot that the perfection of modern European Catholicism and the inauguration of a true science of time were two sides of the same coin, the joint result of a long historical development.

---

qu'ils s'imaginoient de voir être regardé comme le sentiment de l'Eglise.' See also Petavius, *Epistulae* II 13, in *De doctrina*, vol. 3, 318–319.

26 Petavius, *Rationarium, Ad lectorem*, sig. a8r: *Ex tanto chronologiarum numero, quantus ad hanc diem extitit, nullam esse deteriorem, quam quae per summam indignitatem speciosum illud sanctitatis sibi nomen assumpsit. Credo ut ab se inquisitiones hominum, ac iudicia hoc velut sacræ rei et religionis obiectu defenderet. Est autem eiusmodi, qualem neque doctus probare quisquam, nec nisi imperitus laudare possit.*

27 D'Auzoles, *La saincte chronologie.* The work contained a sharp attack on Pétau's *De doctrina*: Stanonik, *Dionysius Petavius*, 67–68. On d'Auzoles' work see the brief comment in Stroumsa, *A New Science*, 81.

7 Time and God

Timekeeping, time measurement, and the care of calendars, Pétau maintains, have always been a matter for priests. In this sense, the time he studies in his work is not merely social time, it is religious time. But why is that so? What is the religious dimension of time? This question has yet to be answered.

Pétau deals with this issue in his dedicatory letter in which he presents his *Doctrina temporum* to Cardinal Richelieu. Right from the outset, he writes about time in strongly theological terms calling it *res sanctissima et sacratissima*.[1] Time, he says, has its origin from the heavens and is for this reason in many ways comparable to God. Like God, time is both the best and the least known thing; everyone is aware it exists, but what it is, even the wisest cannot conceive in their mind, let alone express in words (*ne sapientissimi quidem capere animo nedum verbis explicare possunt*).[2]

Augustine was right, Pétau continues, to express this paradox in his famous words in Book Eleven of the *Confessions*: 'What is time? Provided that no one asks me, I know. If I want to explain it to an enquirer, I do not know.'[3] Yet after that statement, the Church Father continues with sharp and subtle reflections on time only to conclude in Chapter 25 of the same book that he remains absolutely ignorant about the subject of his study.[4]

Furthermore, according to Pétau, time like God contains all things; nothing is outside it. Time precedes everything, manages everything, changes everything. It prescribes to all things their own space in which they act, live and have their being (*sua cuique vel agendi, vel vivendi, vel omnino consistendi spatia praescrib-*

---

[1] Petavius, *De doctrina*, Epistula, vol. 1, sig. a3r: *Res est una sanctissima sacrissimaque tempus*.
[2] Petavius, *De doctrina*, Epistula, vol. 1, sig. a3r: *Nam et originem ducit e coelo; cuius ex motu et conversione potissimum oritur, et excellentem divini numinis in plerisque similitudinem gerit. Ut enim Deus omnium est notissimus idem et ignotissimus: quem et esse nemo non intelligit hominum et cuiusmodi sit nulla potest cogitatio consequi: sic tempus aliquod esse sentiunt omnes, quid sit autem, ne sapientissimi quidem capere animo, nedum verbis explicare possunt*.
[3] Augustine, *Confessions* XI 14, 17: *Quid ergo est tempus? Si nemo ex me quaerat, scio. Si quaerenti explicare velim, nescio.* English translation: Chadwick, *Confessions*, 249.
[4] Petavius, *De doctrina*, Epistula, vol. 1 sig. a3v: *Recte igitur Augustin[us] libro Confession[um] undecimo, postquam multa de natura temporis acute, subtiliterque disseruit, ad extremum nescire se quid sit tempus fatetur quinto et vicesimo capite, cum ita decimo quarto scripsisset* [here follows the citation given in the previous note].

it).⁵ The ancients in their fables, therefore, not only made time a god, but the parent of all the gods.⁶

This ode to time is as intriguing in what Pétau omits as it is in what he says. Time, so much seems clear, is for human beings a holy and sacred reality. This is because in its totalising reality it is experienced as being everywhere and eternal. For this reason, it is both familiar to all and unknowable even to the wisest. It is numinous as much as any deity could be.

The time that is thus described is, of course, what Pétau a few pages later calls 'material' time, the time of the physical and astronomical universe for which he uses many similar expressions even though he avoids in these later pages the theological overtones of the dedicatory letter.⁷ In other words, it is *not* the social time – *formal* time in Pétau's Aristotelian language – in which structure is imposed on time's amorphous infinity in order to make it usable by human societies and cultures.

While it is therefore true that Pétau's initial exposition of time as godlike in its incomprehensibility and inexpressibility anticipates Newton's absolute time, one must not overlook that key for Pétau's argument is the claim that human societies, by means of their religious constitution, move beyond time's apophatic nature and make it knowable.⁸ Otherwise, the science of time he proposes would be impossible:

> Time is such that it can [only] be studied by us insofar as it is adapted to human use and bent to popular notions.⁹

Human beings if left to their own devices, feel a mere sense of awe and bewilderment in the face of infinite time. It is through the assistance of those who shape time into something regular and finite that it becomes available to humanity's benefits and uses. On this, priests and scientists work together.

In light of this overall argument, Pétau's reference to Augustine may seem somewhat backhanded. The Church Father is reported to have retained complete

---

5 Petavius, *De doctrina, Epistula*, vol. 1 sig. a3v: *Ad haec similiter ac Deus, universa continet tempus; rebus omnibus infunditur, abest nusquam, praest omnibus, cuncta moderatur ac mutat, sua cuique vel agendi vel vivendi vel omnino consistendi spatia praescribit.*
6 Petavius, *De doctrina, Epistula*, vol. 1 sig. a3v: *Hinc veteres illi fictis fabulis non solum deos inter adscripsere tempus: sed etiam satorem ac parentem caeterorum esse docuerunt.* On the Kronos-chronos etymology and its background see n. 13 (this chapter) below.
7 Petavius, *De doctrina, Prolegomena* 1, sig. e3v and section 4.1. above.
8 The connection with Newton was emphasised by Wilcox, *The Measure of Time*, 203.
9 Petavius, *De doctrina, Epistula* [no pagination]: *At est eiusmodi tempus: quod sic a nobis considerari potest, quatenus ad humanos accommodatur usus et <ad> populares rationes inflectitur.*

ignorance of time *after* he spent many pages reflecting on it. Is Pétau here implying that Augustine's approach to time was flawed that he, so to speak, barked up the wrong tree? The two certainly disagree on the principal goal of a study of time. Whereas Pétau believes that absolute time, as matter, is inscrutable and that, therefore, any attempt to understand and grasp time in that sense is a fool's errand, Augustine insisted on the necessity of precisely this attempt. Read from Pétau's methodological premises, the Church Father's agonised scrutiny for an answer to the riddle of time can indeed appear evidence that the bishop was in pursuit of an unattainable goal.

Yet Augustine may have erred in a more profound way, according to the French Jesuit. One key claim in *Confessions* XI was that time was created by God alongside the world.[10] For this reason, Augustine maintains, the cheeky question of what God did before he created the cosmos, is beside the point.[11] There was no 'before and after' prior to the creation of the world. Framing a question about God's eternal (and thus pre-cosmic) being in temporal terms is therefore a category mistake. Time is an ineluctable part of the world, but it is, for Augustine, decidedly not part of God or of divine reality.

While Pétau is careful not explicitly to deny Augustine's view which was subsequently accepted by most Christian thinkers, it is nevertheless notable that in his exposition of time's unique character he nowhere even hints at the notion that time was one of God's creatures. As a matter of fact, it is hard to interpret Pétau's use of divine predicates for time as anything but an implicit rejection of the idea that time was created. Time is presented as being on a par with the deity; as Pétau reminds the Cardinal, 'the ancients in their invented myths' went so far as to consider time 'the progenitor (*sator*) and parent' of the other gods.[12]

This is an allusion to the derivation of the term χρόνος (chronos) from the name Kronos which, while etymologically implausible, was popular among Greek authors at least since the Hellenistic age.[13] It then gained new currency

---

10 Book XI of the *Confessions* begins with a lengthy discussion of God's eternity and his creation of time: *Confessiones* XI 1–12 and esp. XI 14,17: *nullo ergo tempore non feceras aliquid, quia ipsum tempus tu feceras*. Cf. Meijering, *Augustin über Schöpfung*, 5–57; Knuuttila, 'Time and Creation'; Flasch, *Was ist Zeit?*, 289–294.
11 Augustine, *Confessions* XI 10, 12–12, 14.
12 Petavius, *De doctrina, Epistula* [no pagination]. See n. 6 (this chapter) above for the full citation.
13 The etymological connection of Kronos with *chronos* is explicitly attested in Plutarch: *De Iside et Osiride* 32, ed. Sieveking, 363 D: οὗτοι δ' εἰσὶν οἱ λέγοντες, ὥσπερ Ἕλληνες Κρόνον ἀλληγοροῦσι τὸν χρόνον. But it is possibly much earlier and may be found in the sixth-century Pherecydes of Syros who lists Chronos as one of three cosmological principles (Diogenes Laërtius 1, 119; Damascius, *De principiis* 124 bis): see Kirk, Raven, and Schofield, *The Presocratic Philosophers*, 56–57.

in the Renaissance as 'Father Time', a figure illustrating 'both the abstract grandeur of a philosophical principle and the malignant voracity of a destructive demon', as Erwin Panofsky put it.[14]

Accordingly, many writers of the period described time 'as a creative or destructive force [...] evoking its paradoxically opposed powers with vivid, contrasting images'. They 'portrayed time as a chameleon being, at once fleeting and destructive, at once the evanescent instant and the ever-recurring cycle'.[15] This in turn led more austere Calvinists, such as Mathieu Brouard (Béroalde: 1520–1576), to embrace a much more theological account of time:

> Only a madman and one with no knowledge of God would ascribe the things that take place in time not to their proper author, God, but to time. For God is the creator of time, as he is of the other things that we see made and unmade with time every day.[16]

Pétau was well aware of Brouard's *Chronicle* as is clear from his critical engagement with some of his (rather eccentric) chronological claims (see above Chapter 4.3). All the more significant is the way in which he positions himself vis-à-vis these recent approaches to time. Despite his reference to the Kronos-chronos etymology, he clearly has no interest in reviving the Renaissance flirtation with a quasi-pagan personification of time. Yet he seems equally far apart from Brouard's disenchanted view of time as a mere creature, and his failure to refer to time as a creature in this context almost certainly implies that he deviates from the traditional Augustinian view on this point. Rather, he considers time as an aspect of the divine and human awareness of time, therefore, as a meeting point of God and humanity.

While the pagans were thus wrong to think of time as 'a' god, they were not wrong to associate it with the sphere of the deity. Human attempts to relate to this reality are therefore an important part of humanity's dealings with the Godhead, for which we use the word religion. As we have seen in the previous chapter, Pétau illustrates this claim through the observation that throughout history priestly hierarchies have played a key role in the management of time.

According to Proclus, the etymology was used in Orphic circles: *In Cratylum* 109, ed. Pasquali, 59, 14 = *fr.* 68 in Kern, *Orphicorum Fragmenta*.
14 Panofsky, 'Father Time', 81.
15 Grafton, *Scaliger*, vol. 2, 344.
16 Beroaldus, *Chronicum* I 1, p. 1–2: *Sic nullus, nisi amens et Dei ignarus, rerum quae cum tempore fiunt causam non in ipsum autorem Deum, sed in tempus contulerit. Temporis enim, ut ceterorum quae cum tempore fieri et evanescere videmus quotidie, Deus opifex est.* English translation: Grafton, *Scaliger*, 344.

We can now see more clearly why he thought this was the case and why to him this was significant. As religion in general mediates a numinous being into the realm of people's experience, so human attempts to structure and measure time bring an otherwise ungraspable reality into a form that can be expressed and used by human societies. Even in this religiously domesticated form, however, time remains fundamentally a divine matter, and its handling by priests is therefore appropriate.

If for Pétau, time thus partakes of God's perfection, this view inevitably has far-reaching consequences for his understanding of God as well. To see this, we have to recall that for Augustine and the later theological tradition, the view that time was created by God coexisted with the assumption that God himself existed eternally outside time.[17] His eternity was therefore not understood as endlessly extended but, in Boethius' famous phrase, as *nunc stans*, an everlasting single moment without duration.[18] God therefore did not literally exist 'before' the world; rather, his eternity is, so to speak, wrapped around the temporal universe. As eternal, God exists apart from the temporal constitution of the cosmos.

In his major theological work, *De theologicis dogmatibus*, Pétau offers a detailed investigation of this question.[19] While the work did not appear until 1644 and thus almost twenty years after *De doctrina*, it is plausible to assume that the author retained his fundamental conviction over these years. In his lengthy discussion of God's eternity, he makes hardly a secret of his opinion that the biblical and Patristic view of the matter is that eternity means eternal duration. Accordingly, he opens the chapter with a citation from Origen according to which eternal is 'what has no beginning in its existence and cannot ever cease to be'.[20] Pétau expounds this by commenting that what is eternal has neither begin-

---

17 See e.g. Augustine, *Confessions* XI 11.13. See also Paul Helm, 'Divine Timeless Eternity', in Ganssle (ed.), *God and Time*, 28–60.
18 Boethius, *De trinitate* IV 64–75, ed. Rand, 20–22: *Quod vero de deo dicitur 'semper est', unum quidem significat, quasi omni praeterito fuerit, omni quoquo modo sit praesenti est, omni futuro erit.* [...] *Semper enim est* [sc. deus], *quoniam 'semper' praesentis est in eo temporis tantumque inter nostrarum rerum praesens, quod est nunc, interest ac divinarum, quod nostrum 'nunc' quasi currens tempus facit et sempiternitatem, divinum vero 'nunc' permanens neque movens sese atque consistens aeternitatem facit.* Among contemporary philosophers the interpretation of Boethius' view remains controversial. See Leftow, 'Boethius on Eternity'.
19 Petavius, *De theologicis dogmatibus* III 3, vol. 1, 189–194.
20 Origen, *De principiis* I 2, 11, ed. Behr, vol. 1, 62, 302–304: *Sempiternum vel aeternum proprie dicitur, quod neque initium ut esset habuit, neque cessare unquam potest esse* [*quod est*]. Cited in Petavius, *De theologicis dogmatibus* III 3, vol. 1, 189. Pétau omits the last two words.

ning nor end, a position he subsequently supports with further Patristic references.²¹

Eternal in this sense, Pétau continues, implies immutability. He cites Richard of Saint Victor to the effect that 'sempiternal' simply means a being without beginning or end whereas 'eternal' signifies *quod caret utroque et omni mutabilitate*.²² God's eternity, then, is not merely never-ending duration but it comes without the changes that are wrought by time on created beings. Pétau adds other divine perfections which hang together with eternity as he understands it, such as absolute power, wisdom and goodness to conclude that eternity can only possibly be ascribed to God in its full and proper sense.²³

Having established this as the predominant view of the tradition, Pétau subsequently deals with the view that God's eternity 'lacks any succession and does not evolve through parts. It does not have a difference of past, present, and future, but exists in a point and in a moment'.²⁴ Pétau, while introducing this view with the famous passage from Book 5 of Boethius' *De consolatione philosophiae*,²⁵ subsequently emphasises that this view originates from Plato and that he will therefore initially explicate it from 'external philosophers'.²⁶

---

21 Petavius, *De theologicis dogmatibus* III 3, vol. 1, 189: *Qua definitione duplicem veluti partem attigit aeternitatis, sive utriusque potius negationem termini, quorum alter initium, alter finis est rei*. Pétau also cites Gregory of Nazianzus, *Oratio* 38, 8 (PG 38, 320 A); Basil, *Adversus Eunomium* I 7 (PG 29, 252 C); and Hebrews 7, 3 which he implies is used by Gregory of Nyssa. He does not give a citation but may have in mind *Contra Eunomium* I 688.

22 Petavius, *De theologicis dogmatibus* III 3, 6, vol. 1, 192: *Triplex dos aeternitatis eius, quae Dei propria est, attingitur. Nam et initium excludit, et successionem, ac mutabilitatem et postremo finem. Quae tria in veram aeternitatem concurrere disertius exponit Richardus Victorinus, cuius verba, quod ad rem magnopere attinent, hic legi oportet. 'Aliud (inquit) sonare videtur aeternum, aliud autem sempiternum. Sempiternum namque esse videtur quod caret initio et fine; aeternum quod caret utroque et omni mutabilitate; et quamvis forsitan neutrum sine altero invenitur, recte tamen inter nominum significationem distinguitur. Quid itaque aliud est aeternitas quam diuturnitas sine initio et fine, et carens omni mutabilitate? Sed qui increatus et sempiternus est caret initio et fine; et cuius status invariabilis est, manet absque omni mutabilitate. In his itaque tribus probatur esse aeternus. Nam haec tria absque ambiguitate dant eternitatem habere et aeternum esse.*' Cf. Richard of Saint Victor, *De trinitate* II 4, ed. Ribaillier, 111, 4–13.

23 Petavius, *De theologicis dogmatibus* III 3, 7–9, vol. 1, 193–194.

24 Petavius, *De theologicis dogmatibus* III 4, 1, vol. 1, 195. The whole fourth chapter (pp. 195–203) is dedicated to this aspect of the topic.

25 Petavius, *De theologicis dogmatibus* III 4, 1, vol. 1, 195. Cf. Boethius, *De consolation philosophiae* V, prose 6. Pétau initially cites the 'definition': *Aeternitas est interminabilis vitae tota simul et perfecta possession* and subsequently gives a long extract from the same section (ed. Rand, 422, 13–424, 38).

26 Petavius, *De theologicis dogmatibus* III 4, 2, vol. 1, 195: *Huic aeternitatis divinae conditioni primum adstipulatores dabo exteros philosophos; orsus a Platone, qui 'id quod semper est, neque*

He fulfils this promise with a long catena of citations from Plotinus, Numenius, Plutarch, Ammonius Hermiae, and Philo before conceding that Christians have followed in their wake.[27] Here Pétau offers, as he says, only a small selection of sources he has to hand *(qui mihi ad manum modo sunt*[28]*)* but which range from Tertullian and the Cappadocians via Augustine to Anselm and Bernard in the early Middle Ages.[29]

Pétau does not say that all these authors were in error, and given the extent of the list and the reputation of the writers chosen, he could hardly expect his readers to think of them as outliers or heretics. Nevertheless, there is but little doubt about his assessment of the different traditions. Whereas in his initial exposition he takes great care to illustrate a consensus among Christian fathers from biblical times onwards, he is now content to give the impression that Christian authors took their cue from an external tradition, namely, Platonism.

In this connection, it is important to take into account that, throughout his work, Pétau goes out of his way to argue that the influence of Platonism on Christianity was overall corrosive.[30] He does not do so here, but the reader is nevertheless invited to make the inference that an understanding of God's eternity according to Plato and his followers can hardly be appropriate from a Christian point of view. He does, moreover, include an incisive passage in which he queries how the notion that God is timeless can be reconciled with the use of the past tense in John 1, 1 *(In principio erat Verbum)*.[31]

---

ortum habet, eodem semper modo habere' dicit. Cf. Plato, *Timaeus* 27d6–28a2. Recent scholarship has pointed out that Plato's understanding of time and eternity is far from straightforward: cf. Sorabji, *Time, Creation, and the Continuum*, 108–112. In fact, Plato's text does not mention time or eternity, but Pétau immediately glosses it with the relevant section from Proclus' *In Timaeum* II, ed. Diehl, 232, 30–31: δεῖ τὸ ἀεὶ ὂν ἐφ' ἑαυτοῦ νοεῖν πόρρω τῆς κατὰ χρόνον μεταβολῆς. He then circles back to Plato, *Timaeus* 37e3–38a8 and ends (this first round) with another quote from Proclus, *In Timaeum* II, ed. Diehl, 239, 2–6.

[27] Petavius, *De theologicis dogmatibus* III 4, 2–3, vol. 1, 195–197. He concludes (III 4, 4, vol. 1, 197): *Haec ex innumeris pauca gentilium, vel minime Christianorum testimonia sufficiunt. Nostrorum autem non minor est copia. qui idem de aeternitate sentientis, definitionem eius illam approbant, quam ex Boëtio supra memoravi. Quae aeternitatem successione omni vacuam differentiaque temporis facit; ac nullas in ea partes constituit; sed simul esse universam in puncto, ac momento consistere docet.*

[28] Petavius, *De theologicis dogmatibus* III 4, 4, vol. 1, 197: *Sed ne longa et odio sit, velut in iudiciis, sic in theologica hac tractatione tot testium appellatio; circumscripta multitudine, certos duntaxat, qui mihi ad manum modo sunt, producemus in medium.*

[29] Petavius, *De theologicis dogmatibus* III 4, 4–9, vol. 1, 197–200.

[30] This has been much discussed in scholarship, most extensively by Hofmann, *Theologie*, 194–224.

[31] Petavius, *De theologicis dogmatibus* III 4, 8, vol. 1, 199.

Let us return to *De doctrina temporum*. There, Pétau is obviously less interested in making theological statements. He does not dwell on the question of what God is or how God is eternal. Instead, his emphasis is firmly on the understanding of time. His alignment of time and God in the opening pages of his dedicatory letter, consequently, has the purpose of characterising time in its absolute aspect as possessing divine attributes. In view of the evidence from his later *Theologica dogmatica*, it is nevertheless hard to doubt that already in *De doctrina* Pétau's underlying view was that God was not outside time but that his eternity was a state of unending duration.

This view throws into even sharper relief Pétau's departure from Augustine's view of time as presented in *Confessions* XI.[32] He disagreed with Augustine on at least two major counts. To begin with, Pétau rejected the Church Father's restriction of time to the realm of created being. Time for Pétau is truly godlike in its infinity, its absoluteness, and its universality. Augustine thus, from Pétau's perspective, underestimated the reach of time and its divine perfection.

That said, this absolute, ungraspable time is not, for Pétau, a worthwhile object of study. On this again, he disagreed with Augustine who sought to understand time in its unstructured, totalising being. This approach, in his view, frustrated Augustine's attempts to make sense of temporal experience and left him with his deep ignorance regarding time which Pétau so strongly emphasises.

Pétau himself proposes a science of time. For this to be possible, time cannot be inscrutable. It is not even enough for it to permit approximate knowledge. As we have seen, the French Jesuit demands strict demonstrability as part of his scientific project. In order to achieve this ambition, he counsels a radical focus on the time that *can* be measured because it has already been appropriated for the use of societies and cultures and is based on the movements of celestial bodies.[33]

In some ways, then, Pétau's project is more modest than Augustine's as it sets aside absolute time as beyond humanity's ken. Yet this epistemic modesty enables a methodological rigour that produces reliable results for those searches that are

---

[32] It should be noted that Pétau had other reasons to show some reserve towards the Bishop of Hippo who was regularly used by the Jesuits' opponents, the Dominicans in the *de auxiliis* controversy and of course the Jansenists. See Fumaroli, 'Temps de croissance', 160–161 and Pétau's remarkable treatise *De Augustini doctrina*.

[33] Intriguingly, Pétau never comments on Augustine's subjective theory of time as *distentio animi* which might have led him to a more positive assessment of Augustine's potential as a theorist of time. After all, for Augustine, the mind (*animus*) does precisely what Pétau seems to ascribe to societies: make time intelligible for human comprehension. Cf. Augustine, *Confessiones* XI 26, 33; Zachhuber, *Time and Soul*, 68–81.

understood to be legitimate. While this reduction of scientific knowledge to the realm of what can be demonstrated seems to prefigure secular principles subsequently adopted in, for example, Kantian epistemologies, Pétau's frame is not secular. After all, the starting point of his work was the claim that, in studying time, we study a sacred (not a secular) reality. Religion and science come together in this exercise because both are human means to grasp and explore divine reality as far as possible and as exactly as possible.

8 Conclusion

This investigation took its starting point from Donald Wilcox' claim that Denis Pétau introduced a radically novel concept of time.[1] Anticipating Newton, Wilcox argued, Pétau used his study of chronology to advance the viewpoint that time properly speaking was one single reality which therefore ought to be measured and calculated in a single, coherent system. This system, according to Wilcox, was the B.C./A.D. scheme, a sequence of years stretching indefinitely into past and future, which came to supersede the previously existing plurality of social times and created for the modern period the sense that humanity had always existed, and continued to exist, in a homogeneous temporal universe in which every moment can be identified as unique and related to any other instant.

The present examination of Pétau's theory in *De doctrina temporum* has confirmed the intuition that a bold and innovative approach to time can be discovered in this work. Pétau's ambition in this regard clearly reached beyond that encountered in his nemesis, Scaliger. As the title to his work promises, Pétau aims at a 'doctrine' or science of time, not merely the 'emendation' of existing calendric systems. Teasing out the character of this 'doctrine' and the understanding of time to which it speaks, turned out to be a rather demanding and complex undertaking, and its eventual results correct earlier scholarship in more than one regard.

Pétau started by distinguishing two aspects of time which he referred to as material and formal time. Of those, the former embraced the notion of infinite and unbounded continuity. It can be studied in the movement of the heavens – whether these are indeed 'infinite' Pétau intriguingly does not address – but it cannot, as such, be understood as time. In order to become intelligible to human beings, material time needs to be brought into order, and this human societies have done from their earliest existence by means of chronological systems in and through which they organised their own communal lives.

Are material and formal time different times, or are they better understood as ways of thinking about and approaching one and the same time? Pétau sends divergent signals in various places, and the exploration of his background also offered pointers in both directions. On the one hand, the Jesuit scholar seemed to draw on concepts distinguishing between absolute time and more limited forms of time. On the other, his nod towards Averroes' distinction of material and formal time would suggest he thinks of two aspects of the same being. What is clear is that he considers 'material' time as a reality beyond humanity's reach, which can-

---

[1] Wilcox, *Measure of Times Past*, 7–9.

ⓐ Open Access. © 2026 the author(s), published by De Gruyter. [CC BY-NC-ND] This work is licensed under the Creative Commons Attribution-NonCommercial-NoDerivatives 4.0 International License.
https://doi.org/10.1515/978-3-11-2223345-009

not be successfully conceptualised, whereas the 'formal' time of human calendars can be known and studied scientifically. It is this latter time which is the object of his own 'science of time'.

The jargon adopted by Pétau throughout his discussion is Aristotelian, but time and again it became evident that his interest in these terms and categories was less indicative of any traditional 'Aristotelianism' and more a launchpad for his own rather original and creative ideas. This was most remarkably the case with the way he inscribed his novel approach to time into the ancient debate about subjective time in Aristotle. He hints at this background merely through his use of the distinction between material and formal time which previously was introduced by Averroes. The Arabic thinker employed this conceptual frame to argue that time existed *potentially* in cosmic motions and was actualised in the mind that had consciousness of time.

In Pétau, this argument for subjective time is changed into a theory of social time. He locates awareness of time in human communities and proposes that their imposition of calendric structures on the vague infinity of cosmic time transforms material into the formal (and intelligible) time which his work will study in detail. Since social time is both cyclical and linear, its study exists in two forms which Pétau calls 'computistics' and 'chronology' of which the former is the comparative investigation of calendars whereas the latter turns to humanity's historical record.

Pétau's (implicit) nod to Averroes' distinction of material and formal time permits him to relate this novel science to Aristotle's definition of time as the 'number of motion or change with regard to before and after'. Formal time is constituted where human societies are conscious of time, but they always have to rely on 'number' and 'motion' which, for Pétau, means they need to employ mathematics and astronomy. This principle applies emphatically to his own science of time which needs to take into account the findings of mathematics and astronomy in order to aspire to the dignity of a science.

With these insights, the investigation turned to the problem of history in Pétau's writing. The co-ordination of the two topics, time and history, could be considered misleading insofar as his treatment of history in the context of *De doctrina* is one part of his comprehensive science of time. Yet this integration of historiography into a broader *doctrina temporum* is in itself significant of Pétau's position towards a heated debate of his time.

This debate concerned the nature of history and the disciplinary character of historiography. Its participants were humanists whose aim was the establishment of an 'art' of historiography, *ars historica*. Pétau's treatment of history in *De doctrina* turned out to be an implicit polemic against this project. Following in the wake of some philosophers, Pétau argued for a limitation of *historia* to purely factual knowledge without its own theoretical apparatus. History is thus reduced to an amorphous mass of information, pure narration of past events. As such, it is 'matter' in need of formal structure which is only provided by chronology.

The precise methodology to be employed in this process is contained in three methodological principles, authority, demonstration, and hypothesis. Chronology is always based on the material, historical record (the 'authority-principle') but can only be considered demonstrably true when it is checked against the 'heavenly clock' by means of mathematical and astronomical calculations ('demonstration principle').

While earlier societies already applied these principles, Pétau clearly thinks that his own work, following in the wake of Scaliger's chronology, which he in equal measure admires and detests, marks a new and in a sense final phase in this history. His scientific chronology is exact and fundamentally infallible due to its methodological rigour and its universal scope. In this connection, he lauds Scaliger's introduction of the Julian Period as a universal timescale against which other chronological systems can be measured but subsequently counsels the use of the B.C./A.D. scheme in its stead as an easily applicable alternative.

As observed above, it is this proposal that has, more than any other aspect of his project, attracted scrutiny as evidence of his advocacy of 'absolute' time in a proto-Newtonian sense. The present investigation has shown the limits of this claim. First of all, his theory of social time means that Pétau interestingly agrees with what scholars have identified as the pre-modern principle according to which time is always the product of a specific society.[2] If there is a proto-Newtonian absolute time in Pétau, it is 'material time' of which there is no real knowledge. Ultimately, we have no access to time in abstraction from a society. This neatly explains his preference of the B.C./A.D. scale over against Scaliger's Julian Period: Pétau opts for the chronological system that is already established in one society, namely, in Western Christendom.

This does not mean, however, that for Pétau this timescale is just one among many. It is unique in its value, but the reason for its exemplary significance, as I have argued, lies in its emergence in the most highly developed society, the same which also (not by coincidence) produced the scientific perfection of chronology. The B.C./A.D. scheme can therefore be called 'absolute', but more in the sense in which Hegelian theologians called Christianity the absolute religion, not in the specific sense of Newton's absolute time.[3] It is the highest possible timescale, the perfection among a range of increasingly valuable chronological systems.

---

2 Feeney, *Caesar's Calendar*, 10.
3 Cf. Troeltsch, *Absoluteness*, ch.1.

Pétau, it appears, held a theory of human progressive development for which the scientific innovations of his age are a prime indicator. Yet this view does not make him an early secularist.[4] Rather, his theory has a religious and ultimately theological frame: progress in science and progress in religion go hand in hand, as evidenced by their convergence in the Gregorian calendar reform. The pope's role in perfecting society's calendric system, Pétau thinks, continues and perfects the work of earlier priests-cum-scientists which have driven advances of the science of time since the dawn of civilisation

Ultimately, the reason for this convergence is the sacredness of time itself. Time is divine, and like God himself, it is partly unknowable, partly known to humanity. There is thus a close analogy between the scientific transformation of infinite and unknowable time into social time that can be understood and studied, and the priestly task of mediating the unknown being of God to humanity at large.

At this point, an intriguing convergence in Pétau's intellectual disposition becomes apparent. A chief concern for him clearly is the limitation and structuring of potentially boundless and infinite reality. Knowledge, to him, is threatened by the overwhelming mass of amorphous information. To understand, then, is to constrain and limit this material. This impetus is evident, first of all, in his insistence that 'material' time needs the structure of human calendars to become intelligible. It is recurrent in his emphasis on the purely factual character of history as the *copia* which has to be shaped into chronology. As we have seen, this leads to what can appear an impoverished and utterly narrow conception of history-as-chronicle, but for Pétau this was clearly the lesser evil (if it was an evil at all) compared with the threat to knowledge from the infinity of the material.

The notion of infinity as inimical to human knowledge and the corresponding idea of the positive value of limit and order is not, of course, Pétau's invention. Rather, with these notions he harks back to ideas that were prominent throughout antiquity. Aristotle's association of infinity with matter (which Pétau affirms) is not an unusual position to hold but rather to the Greeks expressed the truism that perfection lay in an ordered finitude. It is arguable that this ancient consensus was initially breached by the Platonic idea that the Form of the Good was 'beyond being' which, as John Rist explained, meant 'beyond finite being'.[5] The notion that infinity was characteristic of the supreme principle was subsequently

---

4 Pace Klempt, *Säkularisierung*, 86.
5 Rist, *Road*, 23 although with the qualification that 'there is no evidence that Plato [himself already] took this further step'.

incorporated into Christian thought by Gregory of Nyssa and, later, by Dionysius the Areopagite and, in the Middle Ages, in the Franciscan tradition.[6]

It is well known that Pétau was an early campaigner against Platonic influence on Christian theology.[7] One example of this opposition has been discussed in the last chapter – his implicit but clear rejection of the idea that God is outside time. Rarely has it been asked what the underlying motivation was for his hostility to Platonism in Christianity. More research will be needed to give a fully justified response to this question, but at the end of the present investigation, the intriguing possibility emerges that Pétau's anti-Platonism was directly related to his intuition that for knowledge to emerge humanity must leave to one side infinite reality and confine itself to the cognition of being that had already been subject to limitation and structure.

Like Goethe's Faust, the French Jesuit seems persuaded that the Platonic imperative of looking directly into the sun is doomed to fail and that, instead, we should watch 'the waterfall that splits the cliff's broad edge' to observe in 'the rainbow's arch of colour' a metaphor of 'the efforts of mankind':

> Reflect on it, you'll understand precisely:
> We live our lives amongst refracted colour.[8]

---

6 Mühlenberg, *Unendlichkeit*; Schumacher, *Early Franciscan Theology*, ch. 6.
7 Hofmann, *Theologie*, 194–208.
8 Goethe, *Faust: Part II. Act I. Scene I: A pleasant landscape.* English translation: A.S. Kline (2003). Available online at: https://www.poetryintranslation.com/PITBR/German/FaustIIActIScenesItoVII.php. Accessed on 23 July 2025.

Bibliography

NB: Only works referred to elsewhere in this book are listed.

Works by Denis Pétau (Dionysius Petavius)

Synesii Episcopi Cyrenes opera quae extant omnia (ed. and trans.) (Paris: Drovart, 1612).
Iuliani Imperatoris orationes iii panegyricae (ed. and trans.) (La Flèche: Rezé, 1614).
Nicephori Patriarchae Constantinopolitani breviarium historicum de rebus gestis ab obitu Mauricii ad Constantinum usque Copronymum (ed. and trans.) (Paris: Chappelet, 1616).
Themistii cognomento Suadae orationes xix (ed. and trans.) (Paris: Sonnius, 1618).
Epiphanii Constantiae sive Salaminis in Cypro Episcopi opera omnia, 2 vols. (ed. and trans.) (Paris: Sonnius, Morel, and Cramoisy, 1622).
Animadversiones in Epiphanii opus quod Panarium inscribitur, in *Epiphanii [...] opera omnia*, vol. 2 [separate pagination].
Opus de doctrina temporum, 2 vols. (Paris: Cramoisy, 1627).
Tabulae chronologicae regum, dynastarum, urbium, rerum, virorumque illustrium a mundo condito ad annum 4000. Tabula chronologica summorum pontificum, imperatorum, regum, dynastarum a Christo nato ad annum 1628 (Paris: Cramoisy, 1628).
Uranologium sive systema variorum auctorum (Paris: Cramoisy, 1630) = *De doctrina* 1703, vol. 3, 1–220.
Rationarium temporum (Paris: Cramoisy, 1633).
Opus de theologicis dogmatibus, 4 vols. in 5 (Paris: Cramoisy, 1644–1650).
De Augustini doctrina et Tridentina synodo dissertatio (Paris: Cramoisy, 1650).
(Denis Pétav): *La pierre de touche chronologique, contenant la méthode d'examiner la chronologie et en reconnaitre les défauts* [...] (Paris: Cramoisy, 1636).
English translation: Dionysius Petavius: *The History of the World or an Account of Time. [Continued by Others to the Year of Our Lord 1659]* (London: Streater, 1659).
Opus de theologicis dogmatibus, ed. Jean Le Clerc, 6 vols. (Antwerp: Gallet, 1700).
De doctrina temporum, nova editio, ed. Jean Hardouin, 3 vols. (Amsterdam: Gallet, 1703). Cited as *De doctrina* 1703.
Epistularum libri tres, in *De doctrina* 1703, vol. 3, 298–363 [separate pagination].

Ancient texts (prior to 1400)

Aristotle
Ethica Nicomachea, ed. Ingram Bywater (Oxford: Clarendon Press, 1894).
Analytica posteriora, ed. William David Ross (Oxford: Clarendon Press, 1964).
Physica, ed. William David Ross (Oxford: Clarendon Press, 1936).
De anima, ed. William David Ross (Oxford, Clarendon Press, 1961).
Aristotelis Opera cum Averrois commentariis, 12 vols. in 14 (Venice: Junctas, 1562–1574 = Frankfurt am Main: Minerva, 1962).

Augustine
Confessiones, ed. James J. O'Donnell (Oxford: Oxford University Press, 1992).
De civitate dei, eds. Bernard Dombart and Alfons Kalb (Turnhout: Brepols, 1955).

Averroes Cordubensis
Aristotelis de Physico Auditu, in *Aristotelis Opera cum Averrois Commentariis*, vol. IV (Venice: Junctas, 1562 = Frankfurt am Main: Minerva, 1962). Cited as *In physicam*.

Basil of Caesarea
Adversus Eunomium, eds., Louis Doutreleau, George-Matthieu de Durand, and Bernard Sesbouë (Paris: Éditions du Cerf, 1982–1983).

Boethius
De trinitate, ed. Edward Kennard Rand (Cambridge, Mass.: Harvard University Press, 1973), 2–30.
De consolation philosophiae, ed. Edward Kennard Rand (Cambridge, Mass.: Harvard University Press, 1973), 130–435.

Cornelius Celsus
De medicina, ed. Friedrich Marx (Leipzig: Teubner, 1915).

Damascius
De principiis, ed. Charles-Émile Ruelle, 2 vols. (Paris: Klincksieck, 1889 and 1899).

Diodorus Siculus
Bibliotheca historica, eds. Kurt Theodor Fischer and Friedrich Vogel, 5 vols., 3rd edn. (Leipzig: Teubner, 1888–1906).
English translation: Charles Henry Oldfather: *Diodorus Siculus. The Library of History* (Cambridge, Mass.: Harvard University Press, 1950).

Diogenes Laërtius
Vita philosophorum, ed. Tiziano Dorandi (Cambridge: Cambridge University Press, 2013).

Geminos
Elementa astronomiae, ed. Karl Manitius (Leipzig: Teubner, 1898).
English translation: James Evans and J. Leonard Berggren (trans.), *Geminos's Introduction to the Phenomena* (Princeton: Princeton University Press, 2006).

Gregory of Nazianzus
Orationes 38–41, ed. Claudio Moreschini (Paris: Éditions du Cerf, 1990).

Gregory of Nyssa
Contra Eunomium, ed. Werner Jaeger, 2 vols., 2nd edn., *Gregorii Nysseni Opera* I-II (Leiden: Brill, 1960).

Isidore of Seville
Etymologiae, ed. Wallace Martin Lindsay (Oxford: Clarendon Press, 1911).

Julius Africanus
Fragmenta, ed. Martin Wallraff (Berlin: Walter de Gruyter, 2007).

Leo the Great
Epistulae, ed. Jacques Paul Migne, *Patrologia Latina*, vol. 54, 531–1217.

Origen
De principiis, ed. John Behr (Oxford: Oxford University Press, 2017).

Plato
Timaeus, ed. John Burnet, *Platonis Opera*, vol. 4 (Oxford: Clarendon Press, 1902), 17–92 (Stephanus).

Plutarch
De Iside et Osiride, ed. Wilhelm Sieveking (Leipzig: Teubner, 1935).

Proclus
In Platonis Cratylum commentaria, ed. Giorgio Pasquali (Leipzig: Teubner, 1908).
In Platonis Timaeum commentaria, ed. Ernst Diehl, 3 vols. (Leipzig: Teubner, 1903).

Ptolemy
Syntaxis mathematica, ed. Johann Ludvig Heiberg, *Claudii Ptolemaei opera quae exstant omnia*, vol. 1/1 (Leipzig: Teubner, 1908).

Richard of Saint Victor
De trinitate, ed. Jean Ribaillier (Paris: Vrijn, 1958).

Semphronis Asellio
Rerum gestarum libri, ed. Hermann Peter, *Historicorum Romanorum reliquiae*, vol. 1 (Leipzig: Teubner, 1914).

Simplicius [?]
In Aristotelis libros de anima commentaria, ed. Michael Hayduck (Berlin: Reimer, 1882).

Thomas Aquinas
Scriptum super libros sententiarum, eds. Pierre Mandonet and Maria F. Moss (Paris: P. Léthieleux, 1929–1947).
Commentaria in octo libros Physicorum Aristotelis, ed. Leonina XIII., vol. 2 (Rome: Poliglotta, 1884).
Quaestiones disputatae, vol. 2: *De potentia*, ed. Paolo M. Pession, 10th edn. (Rome: Marietti, 1965).

Thucydides
Historiae, eds. Henry Stuart Jones and John Enoch Powell, 2 vols. (Oxford: Clarendon Press, 1942).

Modern literature (from 1400)

Alemanno, Agnese: *Aspetti della cultura teologica nell'Università di Parigi (1604 – 1643): I commenti alla Quaestio II della* Summa Theologiae *di Tommaso d'Aquino (Utrum deus sit)* (PhD Lecce: Conte Editore, 2009).
Annius Viterbiensis, Joannes: *Commentaria Fratris Joannis Annii Viterbiensis super opera diversorum auctorum de antiquitatibus loquentium* (Rome: Eucharius Silber, 1498).
Assmann, Aleida: 'Kulturelle Zeitgestalten', in *Time and History: Proceedings of the 28th International Ludwig Wittgenstein Symposium, Kirchberg am Wechsel, Austria 2005* (Frankfurt am Main: ontos verlag, 2006), 469 – 487.
Barnes, Jonathan: 'Aristotle's Theory of Demonstration', *Phronesis* 14/2 (1969): 123 – 152.
Baur, Ferdinand Christian: *Die christliche Lehre von der Versöhnung in ihrer geschichtlichen Entwicklung von der ältesten Zeit bis auf die neueste* (Tübingen: Osiander, 1838).
Baur, Ferdinand Christian: *Vorlesungen über die christliche Dogmengeschichte*, ed. Ferdinand Friedrich Baur (Leipzig: Fues, 1865).
Bergin, Thomas Goddard and Max Harold Fisch (trans.): *The New Science of Giambattista Vico* (Ithaca, N.Y.: Cornell University Press, 1968).
Bernays, Jacob: *Joseph Justus Scaliger* (Berlin: Hertz, 1855).
Beroaldus, Mattheus: *Chronicum Scripturae Sacrae Constitutum* (Geneva: Chuppinus, 1575).
Bexley, Emmaline: 'Quasi-Absolute Time in Francisco Suárez' Metaphysical Disputations', *Intellectual History Review* 22/1 (2012): 5 – 22.
Bianca, Concetta: 'Gaza, Teodoro', in *Dizionario Biografico degli Itaiani*, vol. 52 (1999). Online at https://www.treccani.it/enciclopedia/teodoro-gaza_(Dizionario-Biografico)/. Accessed on 25 May 2025.
Blum, Paul Richard: *Studies on Early Modern Aristotelianism* (Leiden: Brill, 2012).
Bodin, Jean: *Methodus ad facilem historiarum cognitionem* (Amsterdam: Ravestein, 1650).
Bodin, Jean: *Method for the Easy Comprehension of History*, trans. Beatrice Reynolds (New York: Norton, 1969).
Boyd Davis, Stephen: 'May not Duration Be Represented as Distinctly as Space? Geography and the Visualization of Time in the Early Eighteenth Century', in *Knowing Nature in Early Modern Europe*, ed. David Beck (London: Routledge, 2015), 119 – 137.
Broker, Ralph H.: *The Influence of Bull and Petavius on Cardinal Newman's Theory of the Development of Doctrine* (Rome: Gregoriana, 1938).
Bull, George: *Defensio fidei nicaenae ex scriptis, quae exstant Catholicorum Doctorum, qui intra tria prima Ecclesiae Christianae saecula floruerunt*, 2nd edn. (Oxford: Sheldonian Theatre, 1688 [1685]).
Burnett, Charles: 'Revisiting the 1552 – 1550 and 1562 Aristotle-Averroes Editions', in *Renaissance Averroism and its Aftermath: Arabic philosophy in Early Modern Europe*, eds. Anna Akasoy and Guido Giglioni (Dordercht, etc.: Springer, 2013), 55 – 64.
Calvet, Jean: 'Un confesseur de Saint Vincent de Paul', *Petites Annales de St Vincent de Paul* 4 [no. 41 – 42] (1903): 138 – 146; 166 – 176.
Casaubon, Isaac: *Epistulae*, ed. Theodore Janson ab Almeloveen (Rotterdam: Fritsch & Böhm, 1709).
Chadwick, Henry: *Saint Augustine: Confessions. Translated With an Introduction and Notes* (Oxford: Oxford University Press, 1991).
Chadwick, Owen: *From Bossuet to Newman*, 2nd edn. (Cambridge: Cambridge University Press, 1987 [1957]).
Chytraeus, David: *De lectione historiarum recte instituenda* (Wittenberg: Crato, 1563).

Collegium Conimbricensis: *Commentarii in octo libros Physicorum Aristotelis Stagiritae* (Coimbra: à Mariz, 1592).
Considine, John: 'Isaac Casaubon (1559–1614)', in *Oxford Dictionary of National Bio*graphy. Online at https://doi.org/10.1093/ref:odnb/4851. Accessed on 11 May 2025.
Coope, Ursula: *Time for Aristotle: Physics IV 10–14* (Oxford: Oxford University Press, 2015).
Copernicus, Nicolaus: *De revolutionibus orbium coelestium* (Nuremberg: Petreius, 1543).
Corti, Agustín: *Zeitproblematik bei Martin Heidegger und Augustinus* (Würzburg: Königshausen & Neumann, 2006).
Couzinet, Marie-Dominique: 'History and Philosophy in Francesco Patrizi's Dialoghi della istoria', in *Francesco Patrizi: Philosopher of the Renaissance*, eds. Tomáš Nejeschleba and Paul Richard Blum (Olomouc: Univerzita Palackého, 2014), 62–88.
Coyne, Georde V., Michael A. Hoskin, and Olaf Pedersen (eds.): *Gregorian Reform of the Calendar: Proceedings of the Vatican Conference to Commemorate its 400th Anniversary (1582–1982)*, (Rome: Pontificia Academia Scientiarum, 1983).
Cullmann, Oscar: *Christus und die Zeit: Die urchristliche Zeit- und Geschichtsauffassung*, 3rd edn. (Zürich: EVZ, 1962).
Daniel, Stephen H.: 'Seventeenth-Century Scholastic Treatments of Time', *Journal of the History of Ideas* 42/4 (1981): 587–606.
D'Auzoles Lapeyre, Jacques: *La saincte chronologie du monde divisée en deux parties* (Paris: Alliot, 1632).
De Carvalho, Mário S.: 'Manuel de Góis: The Coimbra Course and the Definition of an Early Jesuit Philosophy', in *Jesuit Philosophy on the Eve of Modernity*, ed. Cristiano Casalini (Leiden: Brill, 2019), 347–372.
De Carvalho, Mário S.: 'The Concept of Time According to the Coimbra Commentaries', in *The Medieval Concept of Time: Studies on the Scholastic Debate and Its Reception in Early Modern Philosophy*, ed. Pasquale Porro (Leiden: Brill, 2001), 353–382.
De Valois, Henri: *Oratio in obitum Dionysii Petavii Societatis Jesu theologi* (Paris, 1653). Reprinted in: Epiphanius of Cyprus, *Opera Omnia*, 2 vols., ed. Dionysius Petavius, new edn. (Cologne: Schrey & Meierl, 1682), sig. **a2r-***a2r.
Detel, Wolfgang: *Subjektive und objektive Zeit: Aristoteles und die moderne Zeit-Theorie* (Berlin: Walter de Gruyter, 2021).
Dorner, Isaak August: *Entwicklungsgeschichte der Lehre von der Person Jesu Christi von den ältesten Zeiten bis auf die neuesten* (Stuttgart: Liesching, 1839).
Edwards, Michael: *Time and the Science of the Soul in Early Modern Philosophy* (Leiden: Brill, 2013).
Feeney, Denis: *Caesar's Calendar: Ancient Time and the Beginnings of History* (Berkeley: University of California Press, 2007).
Flasch, Kurt: *Was ist Zeit? Augustinus von Hippo. Das XI. Buch der Confessiones: Text – Übersetzung – Kommentar* (Frankfurt am Main: Klostermann, 2016).
Foa, Jérémie: 'Le repaire et la bergerie des brebis du Seigneur au milieu da la France: La paysage urbain à Orléans au temps des guerres de Religion', *Histoire urbaine* 41 (2014): 147–168.
Fauchon, Charles: 'La société de Heere ou Saint-Aignan (1615 ?-1624)', *Mémoires de la Société d'agriculture, sciences, belles-lettres et arts d'Orléans*, Fifth Series 18 (1923): 14–28.
Franklin, Julian H.: *Jean Bodin and the Sixteenth-Century Revolution in the Methodology of Law and History* (New York: Columbia University Press, 1963).

Fumaroli, Marc: 'Temps de croissance et temps de corruption: Les deux Antiquités dans l'érudition jésuite française du XVIIe siècle', *XVIIe Siècle: Bulletin de la Société d'étude du XVIIe siècle* 131 (1981): 149–168.

Funck, Johann: *Chronologia* (Wittenberg: Schwertel, 1570).

Galtier, Paul: 'Petau et la preface de son "De Trinitate"', *Revue des Sciences Religieuses* 21 (1931): 462–476.

Ganssle, Gregory E. (ed.): *God and Time: Four Views* (Downers Grove, 2001).

Génébrard, Gilbert: *Chronographiae libri quattuor* (Cologne: Gymnicus, 1581).

Gibbons, Edward: *Memoirs of My Life*, ed. Oliver Farrar Emerson (London: Athenaeum, 1898).

Glocenius, Rudolph: *Lexicon philosophicum* (Frankfurt: Becker, 1613).

Goethe, Johann Wolfgang: *Faust. Part II*, English translation Anthony S. Kline. Online publication. https://www.poetryintranslation.com/PITBR/German/FaustIIActIScenesItoVII.php. Accessed on 23 July 2025.

Grafton, Anthony: 'Joseph Scaliger and Historical Chronology: The Rise and Fall of a Discipline', *History and Theory* 14/2 (1975): 156–185.

Grafton, Anthony: *Defenders of the Text: The Traditions of Scholarship in an Age of Science, 1450–1800* (Cambridge, Mass.: Harvard University Press, 1991).

Grafton, Anthony: *Joseph Scaliger: A Study in the History of Classical Scholarship*, 2 vols. (Oxford: Clarendon Press, 1983 and 1993).

Grafton, Anthony: *What Was History? The Art of History in Early Modern Europe* (Cambridge: Cambridge University Press, 2007).

Grafton, Anthony and Urs B. Leu: '*Chronologia est unica historiae lux:* How Glarean Studied and Taught the Chronology of the Ancient World', in *Heinrich Glarean's Books: The Intellectual World of a Sixteenth-Century Musical Humanist*, eds. Iain Fenlon and Inga Mai Groote (Cambridge: Cambridge University Press, 2013), 248–279.

Hartog, François: *Chronos: The West Confronts Time*, trans. Samuel Ross Gilbert (New York: Columbia University Press, 2022).

Heidegger, Martin: *Sein und Zeit* (Halle: Niemeyer, 1927).

Henderson, Roger D.: 'Vico's View of History', *Philosophia Reformata* 49/2 (1984): 97–111.

Hofmann, Michael: *Theologie, Dogma und Dogmenentwicklung im theologischen Werk Denis Petau's* (Frankfurt am Main: Peter Lang, 1976).

Holt, Mack P.: *The French Wars of Religion, 1562–1629*, 2nd edn. (Cambridge: Cambridge University Press, 2005).

Hornblower, Simon: 'Thucydides, Xenophon, and Lichas: Were the Spartans Excluded from the Olympic Games from 420 to 400 B.C.?', *Phoenix* 54/3–4 (2000): 212–225.

Ideler, Ludwig: *Handbuch der mathematischen und technischen Chronologie*, vol. 2 (Berlin: Rücker, 1826).

Jeck, Udo Reinhold: *Aristoteles contra Augustinum: Zur Frage nach dem Verhältnis von Zeit und Seele bei den antiken Aristoteleskommentaren, im arabischen Aristotelismus und im 13. Jahrhundert* (Amsterdam: Grüner, 1994).

John of Jandun: *Super octo libros Aristotelis De physico auditu subtilissimae quaestiones* (Venice: Hieronymus Scotus, 1586).

Kagan, Donald: *The Peace of Nicias and the Sicilian Expedition* (Ithaca: Cornell University Press, 1981).

Karrer, Leo: *Die historisch-positive Methode des Theologen Dionysius Petavius* (München: Hüber, 1970).

Kecskeméti, Judit: *Fédéric Morel II: Éditeur, traducteur et imprimeur* (Tournhout: Brepols, 2014).

Kern, Otto: *Orphicorum Fragmenta* (Berlin: Weidmann, 1922).

Kirk, Geoffrey S., John E. Raven, and Malcolm Schofield: *The Presocratic Philosophers: A Critical History With a Selection of Texts*, 2nd edn. (Cambridge: Cambridge University Press, 1983).
Klempt, Adalbert: *Die Säkularisierung der universalhistorischen Auffassung: Zum Wandel des Geschichtsdenkens im 16. und 17. Jahrhundert* (Göttingen: Musterschmidt, 1960).
Knuuttilla, Simo: 'Time and Creation in Augustine', in *The Cambridge Companion to Augustine*, eds. David Vincent Meconi and Eleonore Stump (Cambridge: Cambridge University Press, 2014), 81–97.
Koller, Edith: *Strittige Zeiten: Kalenderreform im Alten Reich 1582–1700* (Berlin: Walter de Gruyter, 2014).
Kutsch, Ernst: *Die chronologischen Daten des Ezechielbuches* (Göttingen: Vandenhoeck & Ruprecht, 1985).
Larsson, Gerhard: 'The Chronology of the Pentateuch: A Comparison of the MT and LXX', *Journal of Biblical Literature* 102/3 (1983): 401–409.
Leftow, Brian: 'Boethius on Eternity', *History of Philosophy Quarterly* 7/2 (1990): 123–142.
Levitin, Dmitri: 'From Sacred History to the History of Religion: Paganism, Judaism, and Christianity in European Historiography from Reformation to Enlightenment', *The Historical Journal* 55/4 (2012): 1117–1160.
Levitin, Dmitri: *Ancient Wisdom in the Age of the New Science: Histories of Philosophy in England, c. 1640–1700* (Cambridge: Cambridge University Press, 2015).
Levitin, Dmitri: *The Kingdom of Darkness: Bayle, Newton, and the Emancipation of the European Mind from Philosophy* (Cambridge: Cambridge University Press, 2022).
Losev, Alexandre: '"Astronomy" or "Astrology": A Brief History of an Apparent Confusion', *Journal of Astronomical History and Heritage* 15/1 (2012): 42–46. Online at https://articles.adsabs.harvard.edu//full/2012JAHH...15...42L/0000042.000.html. Accessed on 26 July 2025.
Löwith, Karl: *Meaning in History: The Theological Implications of the Philosophy of History* (Chicago: University of Chicago Press, 1949).
Mandelbrote, Scott: 'Than This Nothing Can Be Plainer: Isaac Newton Reads the Fathers', in *Die Patristik in der frühen Neuzeit*, eds. Günter Frank, Thomas Leinkauf, and Markus Wriedt (Stuttgart – Bad Cannstatt: Frommann – Holzboog, 2006), 277–297.
Mansion, Augustin: 'La théorie aristotélicienne du temps chez les péripateticiens médiévaux', *Revue Neoscolastique de Philosophie* 34 (1934): 275–307.
Meeus, Alexander, Brian Sheridan, and Lisa Irene Hau (eds.): *Diodorus of Sicily: Historiographical Theory and Practice in the 'Bibliotheke'* (Leuven: Peeters, 2018).
Meijering, Eginhard Peter: *Augustin über Schöpfung, Ewigkeit und Zeit: Das elfte Buch der Bekenntnisse* (Leiden: Brill, 1979).
Melanchthon, Philipp: *Chronicon Carionis* (Frankfurt: Zepfellius, 1559).
Momigliano, Arnaldo: 'Vico's Scienza nuova: Roman "Bestioni" and Roman "Eroi"', *History and Theory* 5/1 (1966): 3–23.
Morel, Frédéric (ed.): *Dio Chrysostom, Orationes LXXX* (Paris: Claude Morel, 1604).
Mori, Giuliano: *Historical Truth in Fifteenth Century Italy* (Oxford: Oxford University Press, 2024).
Mosshammer, Alden A.: *The Easter Computus and the Origins of the Christian Era* (Oxford: Oxford University Press, 2008).
Moyer, Ann E.: *Musica Scientia: Musical Scholarship in the Italian Renaissance* (Ithaca: Cornell University Press, 1992).
Mühlenberg, Ekkehard: *Die Unendlichkeit Gottes bei Gregor von Nyssa* (Göttingen: Vandenhoeck & Ruprecht, 1966).

Nelson, Robert: *The Life of Dr. George Bull, Late Bishop of St David's* (London: 1714).
Newton, Isaac: *Scholium on Time, Space, Place, and Motion*, in *Philosophiae Naturalis Principia Mathematica* 1 (1689); trans. Andrew Motte (1729), rev. Florian Cajori (Berkeley: University of California Press, 1934). 6–12. Online at https://plato.stanford.edu/entries/newton-stm/scholium.html Accessed on 8 May 2025.
Nothaft, C. Philipp E.: *Scandalous Error: Calendar Reform and Calendrical Astronomy in Medieval Europe* (Oxford: Oxford University Press, 2018).
Ohashi, Masako: 'Theory and History: An Interpretation of the Paschal Controversy in Bede's *Historia Ecclesiastica*', in *Bède Le Vénérable*, eds. Stéphane Lebecq, Michel Perrin, and Olivier Szerwiniak (Villeneuve d'Ascq: Publications de l'Institut de recherches historiques du Septentrion, 2005). https://doi.org/10.4000/books.irhis.336. Accessed on 1 June 2025.
Oudin, François: 'Denis Petau', in *Memoirs pour server à l'histoire des hommes illustres*, ed. Jean-Pierre Nicéron, vol. 37 (Paris: Briasson, 1737), 81–234.
Panofsky, Erwin: 'Father Time', in Erwin Panofsky: *Studies in Iconology: Humanistic Themes in the Art of the Renaissance* (New York: Harper and Bow, 1962), 95–118.
Patrizi, Francesco: *Della historia dieci dialoghi* (Venice: Arrivabene, 1560).
Pattison, Mark: *Isaac Casaubon, 1559–1614*, 2nd edn. (Oxford: Clarendon Press, 1892).
Pererius, Benedictus: *De communibus omnium rerum naturalium principiis et affectionibus* (Rome: Tramezini, 1576).
Pleins, David J.: *When the Great Abyss Opened: Classic and Contemporary Readings of Noah's Flood* (Oxford: Oxford University Press, 2003).
Pomata, Gianna and Nancy G. Siraisi (eds.): *Historia: Empiricism and Erudition in Early Modern Europe* (Cambridge, Mass.: MIT Press, 2005).
Poole, Robert: *Time's Alteration: Calendar Reform in Early Modern England* (London: UCL Press, 1998).
Porro, Pasquale: *Forme e modelli di durata nel pensiero medievale: L'aevum, il tempo discreto, la categoria 'quando'* (Leuven: Leuven University Press, 1996).
Quantin, Jean-Louis: *Le catholicisme classique et les Pères de l'Église: Un retour aux sources (1669–1713)* (Paris: Institute d'Études Augustiniennes, 1999).
Quantin, Jean-Louis: *The Church of England and Christian Antiquity: The Construction of a Confessional Identity in the 17th Century* (Oxford: Oxford University Press, 2009).
Reiss, Timothy J.: 'Towards the Early Modern Separation of Disciplines: From Philology to Science and History – Joseph Justus Scaliger', *Comparative Literature* 48/2 (1996): 172–179.
Richards, Edward Graham: *Mapping Time: The Calendar and its History* (Oxford: Oxford University Press, 1999).
Ricoeur, Paul: *Temps et récit*, 3 vols. (Paris: Éditions du Seuil, 1983–1985).
Roelli, Philipp: *Latin as a Language of Science and Learning* (Berlin: Walter de Gruyter, 2021).
Rosa, Pietro di: 'Denis Pétau e la cronologia', *Archivum Historicum Societatis Iesu* 57 (1960): 3–54.
Rühl, Franz: 'Die Rechnung nach Jahren vor Christus', *Rheinisches Museum für Philologie*, Neue Folge 61 (1906): 628–629.
Sabrey, Thomas William: *The Person and Work of the Holy Spirit According to the Theories of Denys Petau, SJ., Théodore de Regnon, SJ., and Matthias J. Scheeben* (Washington, PhD CUA, 1952).
Scaliger, Joseph Justus: *De emendatione temporum* (Paris: Mamertus Patissonius, 1583).
Scaliger, Joseph Justus: *Elenchus trihaeresii Nicolai Serarii* (Franeker: Radaeus, 1605).
Scaliger, Joseph Justus: *In Manilii quinque libros Astronomicon commentarius et Castigationes* (Paris: Patissonius, 1579).
Scaliger, Joseph Justus: *Thesaurus temporum*, editio altera (Amsterdam: Janssonius, 1658).

Schlichter, Felix: *Mythology, Chronology, Idolatry: Pagan Antiquity and the Biblical Text in the Scholarly World of Guillaume Bonjour (1670–1714)* (Leiden: Brill, 2025).
Schumacher, Lydia: *Early Franciscan Theology: Between Authority and Innovation* (Cambridge: Cambridge University Press, 2019).
Seifert, Arno: *Cognitio historica: Die Geschichte als Namensgeberin der frühneuzeitlichen Empirie* (Berlin: Dunker & Humblot, 1976).
Snyder, Steven C.: 'Thomas Aquinas and the Reality of Time', *Sapientia* 55 (2000): 371–384.
Sommervogel, Carlos: 'Denis Petau', in Sommervogel, Carlos: *Bibliothèque de la Compagnie de Jésus*, nouvelle edition, eds. Augustin de Backer et al., vol. 6, 2nd edn. (Paris: Picard, 1892), 588–616.
Sorabji, Richard: *Time, Creation, and the Continuum* (London: Duckworth, 1983).
Stanonik, Franz: *Dionysius Petavius: Ein Beitrag zur Gelehrten-Geschichte des XVII. Jahrhunderts* (Graz: Verlag der k. k. Universität, 1876).
Steiner, Benjamin: *Die Ordnung der Geschichte: Historische Tabellenwerke in der Frühen Neuzeit* (Köln: Böhlau, 2008).
Stephenson, F. Richard and Louay J. Fatoohi: 'The Eclipses Recorded by Thucydides', *Historia: Zeitschrift für Alte Geschichte* 50/2 (2001): 245–253.
Stroumsa, Guy: *A New Science: The Discovery of Religion in the Age of Reason* (Cambridge, Mass.: Harvard University Press, 2010).
Suarez, Francisco: *Disputationes metaphysicae*, pars secunda (Venice: Balleoniana, 1751).
Theodore Gaza (trans.): *Aristoteles. De natura animalium* et al. (Venice: Aldine Press, 1504).
Toomer, Gerald James (trans.): *Ptolemy's Almagest* (London: Duckworth, 1984).
Trifogli, Cecilia: 'Averroes' Doctrine of Time and its Reception in the Scholastic Debate', in *The Medieval Concept of Time: Studies on the Scholastic Debate and its Reception in Early Modern Philosophy*, ed. Pasquale Porro (Leiden: Brill, 2001), 57–82.
Trifogli, Cecilia: 'The Unicity of Time in XIIIth Century Natural Philosophy', in *Was ist Philosophie im Mittelalter? Qu'est-ce que la philosophie au moyen âge? What is Philosophy in the Middle Ages? Akten des X. Internationalen Kongresses für Mittelalterliche Philosophie der Société Internationale pour l'Étude de la Philosophie Médiévale, 25. bis 30. August 1997 in Erfurt*, eds. Jan A. Aertsen and Andreas Speer (Berlin: Walter de Gruyter, 1998), 784–790.
Troeltsch, Ernst: *The Absoluteness of Christianity and the History of Religions*, trans. David Reid (Louisville: Westminster John Knox, 2005).
Verbrugghe, Gerald P. and John M. Wickershan (eds.): *Berossos and Manetho Introduced and Translated: Native Traditions in Ancient Mesopotamia and Egypt* (Ann Arbor: University of Michigan Press, 1996).
Vico, Giambattista: *La scienza nuova*, ed. Paolo Rossi (Milan: RCS, 1977).
Völkel, Markus: *'Pyrrhonismus historicus' und 'fides historica': Die Entwicklung der deutschen historischen Methodologie unter dem Gesichtspunkt der historischen Skepsis* (Frankfurt am Main: Lang, 1987).
Vossius, Gerardus: *Ars historica sive de historiae, & historices natura* (Leiden: Mair, 1623).
Weichenhan, Michael: *"Ergo perit coelum": Die Supernova des Jahres 1572 und die Überwindung der aristotelischen Kosmologie* (Stuttgart: Steiner, 2004).
Wilcox, Donald: *The Measure of Times Past: Pre-Newtonian Chronologies and the Rhetoric of Relative Time* (Chicago: University of Chicago Press, 1987).
Zabarella, Jacobus: *De natura logicae* I 2 in Jacobus Zabarella: *Opera Logica* (Venice: Meietus, 1578).

Zachhuber, Johannes: 'Dionysius in the Lutheran Tradition', in *The Oxford Handbook of Dionysius the Areopagite*, eds. Mark Edwards, Dimitrios Pallis, and Georgios Steiris (Oxford: Oxford University Press, 2022), 535–552.

Zachhuber, Johannes: 'The World Soul in Early Christian Thought', in *Platonism and Christianity in Late Ancient Cosmology: God, Soul, Matter*, eds. Ana Schiavoni-Palanciuc and Johannes Zachhuber (Leiden: Brill, 2022), 46–73.

Zachhuber, Johannes: *Time and Soul: From Aristotle to St Augustine* (Berlin: Walter de Gruyter, 2022).

General Index

Aevum   44–45
Albert the Great   43
Ammonius Hermiae   99
annals
– See under *chronicle*
Annius of Viterbo, Annian forgeries   29, 76, n. 75, 90, n. 21
Anselm of Canterbury   99
apodictic, ἀποδεικτικά   70–71
Aristotelianism, Aristotelian tradition   10–12, 38–39, 42–48, 53–54, 58, 62–64, 67, 103
Aristotle   10, 51, 63
– his definition of time   41, 44, 46–50, 55, 103
– Giuntine edition of his works   46
Arianism   88
art, *ars*   10, 53, 62, 65–67
– *ars historica*   10–11, 66–67, 103
– See also under *science*, *scientia*
astronomy, *astrologia*   3, 26–27, 32–33, 40, 49, 51, 52, 54–55, 67, 69, 71, 73, 75, 77, 87, 103
Attic year   29, 33–34
Augustine   1, 97, 99
– *Confessiones*   93–95, 100–101
– *De civitate dei*   43
authority-principle   67–71, 75–77, 82, 104
Averroes (Ibn Rušd)   42, 46–49, 102–103
axioms
– See under *mathematics*

Babylonian, Babylonians   27–28, 31, 89
B.C./A.D. system   6–8, 56, 78–85, 91, 102, 104
Bernard of Clairvaux   99
birth of Jesus Christ, date of   6, 18, 70, 80–82, 91
Bodin, Jean   59–60, 90
Boethius   97–98
Bossuet, Jacques-Bénigne   21
Bourges   17, 44
Brouard (Beroaldus), Matthieu   72–73, 96
Bull, George   21

calendar   3–9, 27–29, 31–36, 51, 55, 77–78, 87–93, 103
Casaubon, Isaac   16, 22
Catholicism
– See under Roman Catholicism
causes   11, 38, 40, 50, 53–54, 62–67, 69–70, 73, 77
Christology   20
chronicle, annals   9–10, 12, 19, 30, 36–37, 59, 65–66, 73–75, 78, 105
Chronicon Carionis   59, 73
chronology
– See under scientific chronology
Chytraeus, David   60–61
Coimbra Commentaries   44
Collège de Clairmont   18
computistics   36, 51–52, 54, 61, 69, 103
Constantine the Great   88
Copernicus, Nicolaus   27
Cornelius Celsus   53
cosmic bodies   3
cosmic soul   47
Council of Nicaea   20, n. 44, 88
creation   1, 9, 19–20, 27, 28, 31, 36, 37, 43–44, 95
Cullmann, Oscar   80

d'Auzoles Lapeyre, Jacques   20, 92
de Gamache, Philippe   16
de Góis, Manuel   44–46, 49, 83
demonstration
– demonstration in Aristotle   53, 64–65,
– demonstration-principle   67–72, 77, 104
Descartes, René   6, 8
Di Rosa, Pietro   15
Diodorus Siculus   34–35
Dionysius Exiguus   28, 80
Duc, Fronton du   17
Duval, André   16

Easter, date of   4, 28, 88–89
eclipse   35, 77
– lunar eclipse   28, 71

- solar eclipse   70–71, 76–77
Egypt, Egyptians   30, 90
equinox   3–4
Essenes   29
eternity   9, 13, 97–100
Eusebius of Caesarea   31, 88
- *Chronicon*   30
Ezekiel   31, 74

Feeney, Denis   7–8
festivals
- See under *religion*
Fumaroli, Marc   22
Funck, Johann   72

Geminos   86–87
Génébrard, Gilbert   72–73
Gassarus, Achilles   59
Gibbon, Edward   25
God   1, 9, 13, 20, 43, 89–90, 93–101
Goethe, Johann Wolfgang   106
Grafton, Anthony   10, 25–26, 59
Gregory of Nyssa   106
Gregory XIII   4, 9, 89
- Gregorian calendar reform   4, 27–28, 89–90, 105

Hartog, François   81
Hebrew Bible, chronology of   29
Hegelianism   104
Heidegger, Martin   1–2,
Herodotus   30
historiography   9–13, 26, 57, 59, 80–81, 83–84, 103
history
- as *rerum praeteritarum narratio*   60
- as *narratio vera*   60
- *historia* in Pétau   11, 54, 61–67, 84, 103
- *historicé* (in Vossius)   65–66
Hofmann, Michael   14–15, 18, 22
humanism, humanists   4, 10, 15, 16, 59, 62, 63, 84, 103
hypothesis-principle   67, 69–70, 72, 78–79, 81–84, 104

Ideler, Ludwig   25, 33
immutability   98

Incarnation   8, 20, 80, 91
Isidore of Seville   60

Jesuits   17, 22, 24, 32–33, 44, 91, 100 with n. 32
John of Jandun   43–45
Julian calendar   3–4, 9, 90
Julian period   28, 35, 78–84, 104
Julius Africanus   30–31
Julius Caesar   3, 88, 90

Klempt, Adalbert   8, 22, 80–81, 91
Kronos-Chronos   95–96

La Flèche   17
Le Clerc, Jean   21
Löwith, Karl   80, 91
lunar cycle   34–35, 41, 87

Manetho   30
Marcus Manilius   26
mathematics   3, 41, 49, 52–55, 61, 67, 69, 73, 75, 103
- mathematical axioms, postulates   68, 72, 78–79, 81–82
Morel, Frédéric II   16–17
music   40–41

Nabonassar   27
Nabopollassar   31
Nancy   17
Newton, Isaac   5–7, 83, 94, 102
Numenius   99

Orléans   15–18
Oudin, François   14–15, 17, 91

Panofsky, Erwin   96
Patrizi, Francesco   64–66, 73
Peloponnesian War   35, 70–73, 76, 79
Pétau, Françoise   15
Pétau, Jerôme   15
Pétau, Paul   15
Philo of Alexandria   29, 99
physics   1, 40, 42, 44, 49
Plato, Platonism   66, 98–99, 105–106
Plotinus   99

General Index — **119**

Plutarch   99
Pont-à-Mousson   17
postulates
– See under *mathematics*
priests   9, 89–90, 93–94, 97, 105
primum mobile, celestial sphere   42–43, 45
Protestantism, Protestants   5, 13, 15, 17, 24, 33
Ps-Dionysius the Areopagite   106
Ptolemy   28

Rationalist and empiricist schools of medicine   53
Reims   17
religion   13, 86, 89–92, 96–97
– absolute religion   104
– history of religions   8, 91–92
– religious festivals   2, 87–88
– science and religion   12, 92, 101, 105
Renaissance   96
Richard of Saint Victor   98
Richelieu, Cardinal (Armand Jean du Plessis)   9, 85, 89, 93
Roman Catholicism, Tridentine Catholicism   5, 8, 15, 22, 91–92

sacrifice   87
Salmanassar (Shalmaneser V)   27
Scaliger, Joseph Justus   5, 12–13, 15, 17, 22, 24–37, 53 72, 76–80, 82–85, 90, 102, 104
– *De emendatione temporum*   26–27, 29, 33, 36, 59
scepticism, sceptics   60, 64
science, *scientia*   23, 36, 40–42, 52, 59, 67, 69, 75, 81, 90, 105
– *ars* and *scientia*   11, 63–65, 67
– *scientia temporum*   8, 38, 49–54, 56–57, 61, 78, 86–89, 91, 94, 100–103
– For science and religion see under *religion*
scientific chronology   3, 9–10, 12, 19, 24–26, 32–34, 57, 72–73, 83–84, 90–91

Scripture   12, 30–31, 74, n. 70
secularisation/secularity   23, 80, 105
Seifert, Arno   54
Septuagint, chronology of   29
Sicilian Expedition   34
solar year   3–4, 28, 77, 87

temporality   2, 52, 55, 80, 83, 86
Theodore Gaza   63–65
Thomas Aquinas   43, 47
Thucydides   12, 68, 70–71, 76–77
Time
– absolute time   5–7, 81, 83, 94–95, 100, 102, 104
– created by God   43, 95, 97, 100
– imagined time (*tempus Imaginarium*)   45–46, 49
– material and formal time   38–40, 42, 46–50, 78, 83, 94, 102–103
– proleptic time   30
– relative time   5–6
– sacred reality   89, 94, 101, 105
– social time   3–9, 49–50, 55, 78, 81–82, 84–92103–105
– subjective and objective time   1–3, 47–49, 55, 86, 100, n. 33, 103
– See also under *Kronos-chronos*
Trinity   20

Vico, Giambattista   25, 74
Vos, Gerrit (Gerardus Vossius)   24, 65–66, 73–74

Wilcox, Donald   5–9, 14, 24, 81, 83, 102

Xenophon   68, 71–72

Ysambert, Nicolas   16

Zabarella, Giacomo   53, 62–63, 65–66, 73

Index of Passages

Aristotle
– *Ethica Nicomachea*
  – VI 4   10
– *Analytica posteriora*
  – I   10, 53
  – I 10   51
– *Physica*
  – IV 11   41, 48, 55
  – IV 13   39
  – IV 14   46
– *De anima*
  – I 1   63
Augustine
– *Confessiones*
  – XI 1–12   95
  – XI 14,17   95
  – XI 23,29   43
  – XI 26,33   1, 100
– *De civitate dei*
  – XII 16   43
Averroes Cordubensis
– *In physicam*
  – IV, t. c. 109   48
  – IV, t. c. 131   47–48
  – IV, t. c. 132   42

Basil of Caesarea
– *Adversus Eunomium*
  – I 7   98
Biblia sacra
– *2 Regum*
  – 18, 9   27
– *Ezechiel*
  – 1, 1   31, 74
– *Sec. Ioannem*
  – 1, 1   99
Boethius
– *De trinitate*
  – IV 64–75   97
– *De consolation philosophiae*
  – V, prose 6   98

Cornelius Celsus
– *De medicina*
  – I, prooem.   53

Damascius
– *De principiis*
  – 124 bis   95
Diodorus Siculus
– *Bibliotheca historica*
  – XIII 2,4   34–35
Diogenes Laërtius
– *Vita philosophorum*
  – 1, 119   95

Geminos
– *Elementa astronomiae*
  – VII 6–7   87–88
Gregory of Nazianzus
– *Orationes*
  – 38, 8   98
Gregory of Nyssa
– *Contra Eunomium*
  – I 688   98

Isidore of Seville
– *Etymologies*
  – I 41.1   60, 63

Julius Africanus
– *Fragmenta*
  – F 43   30
  – F 46   30

Leo the Great
– *Epistulae*
  – 121, 1   88

Origen
– *De principiis*
  – I 2, 11   97

Plato
- *Timaeus*
  - 26d6–28a   99
  - 37e3–38a8   99
Plutarch
- *Alcibiades*
  - 20   34
- *De Iside et Osiride*
  - 32   95
Proclus
- *In Cratylum*
  - 109   96
- *In Timaeum*, ed. Ernst Diehl
  - Vol. 1, p. 232, 30–31   99
  - Vol. 1, p. 239, 2–6   99
Ptolemy
- *Syntaxis mathematica*
  - III 7   28

Richard of Saint Victor
- *De trinitate*
  - II 4   98

Semphronis Asellio
- *Rerum gestarum libri*
  - 1, fr. 1   66
Simplicius [?]
- *In de anima*, ed. Michael Hayduck
  - p. 7, 27–28   63

Thomas Aquinas
- *Commentaria in libros sententiarum*
  - II d. 12, qu. 1, art. 5   44
- *In Physicam*
  - IV, lect. 23, n. 5   47
- *De potentia*
  - Qu. 5, art. 5   44
Thucydides
- *Historiae*
  - II 28, 1   72
  - IV 52, 1   72
  - V 49–50   72
  - VII 50, 4   72

The following volumes have been published in this series:

Volume 2
Detel, Wolfgang. *Subjektive und objektive Zeit: Aristoteles und die moderne Zeit-Theorie.* Berlin/Boston: De Gruyter, 2021.

Volume 3
Singer, P. N. *Time for the Ancients: Measurement, Theory, Experience.* Berlin/Boston: De Gruyter, 2022.

Volume 4
Gertzen, Thomas L. *Aber die Zeit fürchtet die Pyramiden: Die Wissenschaften vom Alten Orient und die zeitliche Dimension von Kulturgeschichte.* Berlin/Boston: De Gruyter, 2022.

Volume 6
Zachhuber, Johannes. *Time and Soul: From Aristotle to St. Augustine.* Berlin/Boston: De Gruyter, 2022.

Volume 7
Golitsis, Pantelis. *Damascius' Philosophy of Time.* Berlin/Boston: De Gruyter, 2023.

Volume 8
Defaux, Olivier. *La Table des rois: Contribution à l'histoire textuelle des ›Tables faciles‹ de Ptolémée.* Berlin/Boston: De Gruyter, 2023.

Volume 9
Fischer, Julia (ed.). *Zwiegespräche über die Zeit: Dialoge in der Berlin-Brandenburgischen Akademie der Wissenschaften aus Anlass des sechzigsten Geburtstags von Christoph Markschies.* Berlin/Boston: De Gruyter, 2024.

Volume 10
Walter, Anke (ed.). *The Temporality of Festivals: Approaches to Festive Time in Ancient Babylon, Greece, Rome, and Medieval China.* Berlin/Boston: De Gruyter, 2024.

Volume 12
Sieroka, Norman. *Zeit-Hören: Erfahrungen, Taktungen, Musik.* Berlin/Boston: De Gruyter, 2024.

Volume 13
Birk, Ralph/Coulon, Laurent (ed.). *The Thebaid in Times of Crisis: Revolt and Response in Ptolemaic Egypt*. Berlin/Boston: De Gruyter, 2025.

Volume 14
Pallavidini, Marta. *(A)synchronic (Re)actions: Crises and Their Perception in Hittite History*. Berlin/Boston: De Gruyter, 2025.

Volume 15
Bech Nosch, Marie-Louise. *Time and Textiles in Ancient Greece*. Berlin/Boston: De Gruyter, 2026.

Volume 16
Klinger, Jörg. *Das Erfassen von Zeit im Kontext der Vergangenheit*. Berlin/Boston: De Gruyter, 2026.

www.ingramcontent.com/pod-product-compliance
Lightning Source LLC
Chambersburg PA
CBHW051543230426
43669CB00015B/2707